AF732731

Environmental Journalism

Environmental journalism is an increasingly significant area for study within the broader field of journalism studies. It connects the concerns of politics, science, business, culture and the natural world whilst also exploring the boundaries between the local, regional and global. A central and typical focus for its concerns are the global summits convened to share scientific knowledge about global warming and to formulate policies to mitigate its consequences in particular locales. But reporting environmental change creates difficulties for journalists who are often ill equipped to resolve the uncertainties in the disputed scientific accounts of climate change.

This research-based collection focuses on aspects of environmental journalism in Australia, France, Norway, Sweden, the UK and the USA. Contributors present case studies of media reporting of the environment, and explore considerations of objectivity and advocacy in journalistic coverage of the environment and climate change.

This book was originally published as a special issue of *Journalism Studies.*

Henrik Bødker is Associate Professor in the Department of Aesthetics and Communication – Media Science at Aarhus University, Denmark.

Irene Neverla is Programme Director in the Institute for Journalism and Communication at the University of Hamburg, Germany.

Journalism Studies: Theory and Practice

Series editor: Bob Franklin, Cardiff School of Journalism, Media and Cultural Studies, Cardiff University, UK

The journal *Journalism Studies* was established at the turn of the new millennium by Bob Franklin. It was launched in the context of a burgeoning interest in the scholarly study of journalism and an expansive global community of journalism scholars and researchers. The ambition was to provide a forum for the critical discussion and study of journalism as a subject of intellectual inquiry but also an arena of professional practice. Previously, the study of journalism in the UK and much of Europe was a fairly marginal branch of the larger disciplines of media, communication and cultural studies; only a handful of Universities offered degree programmes in the subject. *Journalism Studies* has flourished and succeeded in providing the intended public space for discussion of research on key issues within the field, to the point where in 2007 a sister journal, *Journalism Practice,* was launched to enable an enhanced focus on practice-based issues, as well as foregrounding studies of journalism education, training and professional concerns. Both journals are among the leading ranked journals within the field and publish six issues annually, in electronic and print formats. More recently, 2013 witnessed the launch of a further companion journal *Digital Journalism* to provide a site for scholarly discussion, analysis and responses to the wide ranging implications of digital technologies for the practice and study of journalism. From the outset, the publication of themed issues has been a commitment for all journals. Their purpose is first, to focus on highly significant or neglected areas of the field; second, to facilitate discussion and analysis of important and topical policy issues; and third, to offer readers an especially high quality and closely focused set of essays, analyses and discussions.

The *Journalism Studies: Theory and Practice* book series draws on a wide range of these themed issues from all journals and thereby extends the critical and public forum provided by them. The Editor of the journals works closely with guest editors to ensure that the books achieve relevance for readers and the highest standards of research rigour and academic excellence. The series makes a significant contribution to the field of journalism studies by inviting distinguished scholars, academics and journalism practitioners to discuss and debate the central concerns within the field. It also reaches a wider readership of scholars, students and practitioners across the social sciences, humanities and communication arts, encouraging them to engage critically with, but also to interrogate, the specialist scholarly studies of journalism which this series provides.

Available titles in the series:

Mapping the Magazine: Comparative Studies in Magazine Journalism
Edited by Tim Holmes

The Future of Newspapers
Edited by Bob Franklin

Language and Journalism
Edited by John Richardson

The Future of Journalism
Edited by Bob Franklin

Exploration in Global Media Ethics
Edited by Muhammad Ayish and Shakuntala Rao

Foreign Correspondence
Edited by John Maxwell Hamilton and Regina G. Lawrence

How Journalism Uses History
Edited by Martin Conboy

Lifestyle Journalism
Edited by Folker Hanusch

Environmental Journalism
Edited by Henrik Bødker and Irene Neverla

Online Reporting of Elections
Edited by Einar Thorsen

The Future of Journalism: Developments and Debates
Edited by Bob Franklin

Environmental Journalism

Edited by
Henrik Bødker and Irene Neverla

Routledge
Taylor & Francis Group
LONDON AND NEW YORK

First published 2013
by Routledge
2 Park Square, Milton Park, Abingdon, Oxon, OX14 4RN

Simultaneously published in the USA and Canada
by Routledge
711 Third Avenue, New York, NY 10017

Routledge is an imprint of the Taylor & Francis Group, an informa business

© 2013 Taylor & Francis

This book is a reproduction of *Journalism Studies*, volume 13, issue 2. The Publisher requests to those authors who may be citing this book to state, also, the bibliographical details of the special issue on which the book was based.

All rights reserved. No part of this book may be reprinted or reproduced or utilised in any form or by any electronic, mechanical, or other means, now known or hereafter invented, including photocopying and recording, or in any information storage or retrieval system, without permission in writing from the publishers.

Trademark notice: Product or corporate names may be trademarks or registered trademarks, and are used only for identification and explanation without intent to infringe.

British Library Cataloguing in Publication Data
A catalogue record for this book is available from the British Library

ISBN13: 978-0-415-82749-2

Typeset in Helvetica
by Taylor & Francis Books

Publisher's Note
The publisher would like to make readers aware that the chapters in this book may be referred to as articles as they are identical to the articles published in the special issue. The publisher accepts responsibility for any inconsistencies that may have arisen in the course of preparing this volume for print.

Contents

Citation Information

The chapters in this book were originally published in *Journalism Studies*, volume 13, issue 2 (April 2012). When citing this material, please use the original page numbering for each article, as follows:

Preface
Original title in journal special issue: EDITORIAL NOTE
Bob Franklin
Journalism Studies, volume 13, issue 2 (April 2012) pp. 151

Chapter 1
Introduction: Environmental Journalism
Henrik Bødker and Irene Neverla
Journalism Studies, volume 13, issue 2 (April 2012) pp. 152-156

Chapter 2
Climate Change Controversies in French Mass Media 1990-2010
Stefan Cihan Aykut, Jean-Baptiste Comby, and Hélène Guillemot
Journalism Studies, volume 13, issue 2 (April 2012) pp. 157-174

Chapter 3
Framing Climate Change: A study of US and Swedish press coverage of global warming
Adam Shehata and David Nicolas Hopmann
Journalism Studies, volume 13, issue 2 (April 2012) pp. 175-192

Chapter 4
Media Representations of Climate Change in the Argentinean Press
María Teresa Mercado
Journalism Studies, volume 13, issue 2 (April 2012) pp. 193-209

Chapter 5
Participatory Politics, Environmental Journalism and Newspaper Campaigns
Anita Howarth
Journalism Studies, volume 13, issue 2 (April 2012) pp. 210-225

Chapter 6
Environmental Journalism in Bangladesh: Active social agency

Jahnnabi Das
Journalism Studies, volume 13, issue 2 (April 2012) pp. 226-242

Chapter 7
Playing the Media Game: The relative (in)visibility of coal industry interests in media reporting of coal as a climate change issue in Australia
Wendy Bacon and Chris Nash
Journalism Studies, volume 13, issue 2 (April 2012) pp. 243-258

Chapter 8
Setting the Agenda on Environmental News in Norway: NGOs and newspapers
Roy Krøvel
Journalism Studies, volume 13, issue 2 (April 2012) pp. 259-276

Chapter 9
Talking Points Ammo: The use of neoliberal think tank fantasy themes to delegitimise scientific knowledge of climate change in Australian newspapers
Elaine McKewon
Journalism Studies, volume 13, issue 2 (April 2012) pp. 277-297

NOTES ON CONTRIBUTORS

Stefan Cihan Aykut is researcher and PhD student at Centre Alexandre Koyré, associated researcher at the German-French Institute (DFI) and lecturer at Sciences Po Paris. His research interests include international climate governance, comparative analysis of climate change discourses and politics in France and Germany, and links between energy and climate politics. He participated in two United Nations climate conferences as an observer and published several papers and reports on the subject. Recent publications include "France and Germany in the Post-Copenhagen Process" (La France et l'Allemagne dans le cadre du processus post-Copenhague) in *Annuaire français des relations internationales*. He is responsible for a research project on climate change controversies financed by GIS Climat Environnement Société.

Wendy Bacon is a professor of journalism based at the Australian Centre for Independent Journalism at the University of Technology, Sydney (UTS). A well-known Australian investigative journalist and non-practising lawyer, she publishes with *The Sydney Morning Herald*, Crikey.com and Reportage Online. Her research has, among other places, been published in *Australian Journalism Review* and *Pacific Journalism Review*.

Henrik Bødker is Director of the Centre for University Studies in Journalism and Associate Professor at the Department of Aesthetics and Communication, Aarhus University, Denmark. His research and teaching interests span the fields of popular culture (music), media and journalism. He is currently working on a project focusing on the roles of commentary within journalism. He has contributed to various journals and anthologies.

Jean-Baptiste Comby is Associate Professor at the French Press Institute, University of Paris 2. He teaches media sociology and works as a researcher at the CARISM (Center for Interdisciplinary Researches and Analysis on Medias). His publications include book chapters and articles on the social construction of climate change as a public problem. He also leads research and publishes on how people deal with environmental issues. His last publication in English deals with a different topic: the way people use the Internet to get informed about current affairs (with Josiane Jouet and Thierry Vedel, "Political Information and Interpersonal Conversations in a Multimedia Environment: a quantitative and qualitative examination of information practices in France", *European Journal of Communication* 26(4), 2011).

Jahnnabi Das is a PhD candidate and teaches journalism and communication at the University of Technology, Sydney (UTS). She is a former newspaper journalist from Bangladesh, and has had professional training in Europe. Her area of interest in research is professional practice in journalism. She has written on professional issues in freelance journalism and environmental reporting practices.

Hélène Guillemot, PhD, is researcher at Centre Alexandre Koyre and the Centre National de la Recherche Scientifique in Paris. She studies scientific practices of climate modelling and the relationship between scientific research and expertise in climate science communities. Her recent publications include "Connections Between Climate Simulations and Observation in Climate Computer Modeling. Scientist's practices and 'bottom-up epistemology' lessons", *Studies in History and Philosophy of Modern Physics* 41(3), 2010 and "Climate Change: scientific dynamics, expertise and political challenges" (with A. Dahan) in *Science and Sovereignty* (edited by C. Paradeise, Routledge). She is responsible for a research project gathering social scientists and climatologists on the questions of confidence in global warming.

David Nicolas Hopmann, PhD, is Assistant Professor in Media and Politics at the Centre for Journalism, Department of Political Science, University of Southern Denmark. His research deals with political communication in the media and in interpersonal settings, which has been published in leading journals such as the *Journal of Communication, Political Communication* and *European Journal of Political Research*.

Anita Howarth worked as a journalist for 10 years and now lectures in political communication, advocacy and campaigning at Kingston University, London. She has a BA in history from the University of Natal, a Masters in politics from Birkbeck, and her PhD from LSE analyses the relationship between media and public policy on genetically modified food. She has published on environmental politics and journalism, media campaigns and on political communication in South Africa.

Roy Krøvel has a PhD in History. His dissertation was on the relationship between the media and guerrilla organizations and indigenous peoples in Mexico and Central America. He is also a civil engineer specializing in the environment and risk analysis. Krøvel is currently Associate Professor in Journalism at Oslo and Akershus University College in Norway. Outside academia Krøvel has published several books on journalism and conflict. He is currently project manager for the Master in Intercultural Communication in cooperation with Universidad de las Regiones Autónomas de la Costa Caribe Nicaragüense in Nicaragua.

Elaine McKewon is a PhD candidate at the University of Technology, Sydney. Her study examines the reconstruction of the scientific consensus on climate change as a "scientific debate" in the Australian news media.

María Teresa Mercado, PhD, is Associate Professor at the Department of Communication and Journalism at the CEU UCH University in Valencia, Spain. She teaches Environmental Journalism, among others subjects. She is Deputy Director of the Institute for Environmental Studies and member of the Research Group for Disability and Communication (GIDYC) at CEU UCH. Her publications include articles on environmental, disability and cultural heritage issues in news coverage of Spanish media. She is recently co-author of two book chapters in English: "Rare Diseases: information hierarchy and the reader's participation in general and specialised media" and "News Values, Frames and Rare Diseases: novelty, curiosity and proximity", in *Communication Strategies and Challenges for Rare Diseases: medical research as a referent* (edited by CIBERER, Madrid, 2011). She spent six months in a research stay at the Austral University in Buenos Aires.

Chris Nash is Professor and Head of the Journalism programme at Monash University. His current research and teaching interests are in journalism as intellectual and social practice, particularly with respect to the built and natural environment, and issues in sustainability. Professionally, he has worked as a journalist in radio, television and documentary film.

Irene Neverla, Prof. Dr., Chair of Journalism and Communications in the Faculty of Social Sciences and Economy at the University of Hamburg, Germany. Her research work and publications are on the public discourse (seen both from journalism theory and from audience research), environmental communication (with a special focus on the public discourse on climate change), visual communication and press photography, European and other transnational public spheres. She has published in *European Societies*; in *Difference and Democracy. Exploring potentials in Europe and beyond* (edited by Raube and Sattler, Frankfurt am Main: Campus, 2011); and has co-edited with Mike Schaefer *Das Medien-Klima. Fragen und Befunde der kommunikationswissenschaftlichen Klimaforschung* (Wiesbaden: Verlag Sozialwissenschaften, 2012).

Adam Shehata is Assistant Professor in Media and Communication at Mid Sweden University. His research focuses on news coverage, media effects and political engagement; this work has been published in journals such as *Communication Research*, *Journal of Communication* and *Political Communication*.

PREFACE

Environmental journalism is an increasingly significant area for study within the broader field of journalism studies. Henrik Bødker (Aarhus University, Denmark) and Irene Neverla (University of Hamburg, Germany), who have edited the collection of essays published here, suggest that this significance reflects the fact that environmental journalism connects the concerns of politics, science, business, culture and the natural world, but also explores the boundaries between the local, regional and global. A central and typical focus for its concerns are the global summits convened to share scientific knowledge about global warming and to formulate policies to mitigate its consequences in particular locales. But reporting environmental change creates difficulties for journalists who are often ill equipped to resolve the uncertainties in the disputed scientific accounts of climate change. Further problems emerge from the need to balance the journalistic requirement for objective coverage when confronted with the prospect of famines and flood which might call for a journalistic approach tempered by advocacy.

These research-based studies which focus on aspects of environmental journalism in Australia, France, Norway, Sweden, the United Kingdom and the United States, fall into three broad sections. The first three essays present case studies of media reporting of the environment, while the subsequent two contributions explore considerations of objectivity and advocacy in journalistic coverage of the environment and climate change. The final three essays offer intriguing accounts of how substantial, but non-journalistic actors and agencies, (including non-governmental organisations, business and think tanks), help to shape the media and public agenda on climate change but typically in ways which too often elude public notice.

I am very grateful to Henrik Bødker and Irene Neverla for issuing the call for papers and gathering together this distinguished group of scholars whose essays are published here. I am also delighted to have an opportunity to thank Dr Neil Gavin (Liverpool University, UK) and Dr Libby Lestor (University of Tasmania, Australia) for agreeing to read, comment and offer advice on earlier drafts of the papers.

Finally, I hope you will enjoy reading these essays which at times provide rather unsettling as well as enlightening accounts.

Bob Franklin
Editor

INTRODUCTION
Environmental journalism

Henrik Bødker and **Irene Neverla**

There are many reasons for considering environmental journalism as an academically significant, challenging and valuable focus for a themed issue of *Journalism Studies*. Journalism about the environment and climate change sits at a complex of intersections between politics, business, science, nature and culture, in between, the individual and the common but also in between the local, regional and global levels. While the environment largely appears as local and tangible, the related issues of climate change are global and intangible. This means, that in some locations, associated risks are contemporary and very real, while in others, perhaps most, such risks are remote in both time and space and thus only knowable through various representations. Such differences in perception are, however, not only linked to geographical location but also, at least to some extent, linked to levels of income and technological development (two aspects, which together may allow for more sustained mitigation measures). While there is a broad scientific consensus on the anthropogenic contributions to global warming, scientific knowledge on the regional impacts of global warming is still somewhat uncertain. This, along with the vast differences in perceptions among individuals and in between countries, make global political attempts at mitigation fraught with economical, scientific, developmental, geographical, social and cultural fault lines—as COP 15 in Copenhagen illustrated neatly.

Journalism was an intimate part of establishing and negotiating such fault lines and a focus on conflict is an established journalistic approach to complex issues. Away from global summits, however, climate change is more difficult for journalists to handle since this issue breaches some of journalism's established foci and demarcations. Firstly, the temporal aspect makes climate change difficult to handle journalistically; apart from melting glaciers and extreme weather events, the issue of global warming is difficult to make concrete, and thus the subject of news. This is somewhat linked to the fact that environmental issues, at least in some measure, cut across established editorial boundaries such as politics, business, technology, nature, culture and consumer issues. Finally, since climate change is threatening to inflict great upheaval through famines, and migration waves, the question arises of whether traditional professional dichotomies between "objective" and "balanced" versus "advocacy" journalism are adequate in the face of such social and environmental challenges. This question is exacerbated by the fact that this is an area of reporting that is heavily dependent on journalistic translations of scientific knowledge. Taken together these aspects make environmental journalism an area ripe for clashes of various sorts.

This, as well as a broadly conceived urgency of the issue, has increasingly made journalistic coverage of climate change the focus of media, communication and journalism studies. By studying something that does not quite fit, established practices are made more visible; and emergent practices may be glimpsed. An issue with no clearly defined boundaries may arguably push communication practices to relate to place and space in new ways. Thus, being a window into processes of globalisation it may not only reveal something about new ways of communicating; it may also reveal aspects of a new, emerging world order. Yet, in relation to journalism, it is still important to point out that this most often is written from and to specific places and therefore inflected by local political, economic, geographical, media-related, professional and cultural factors in various ways.

This has come out of a number of studies focusing on domestication and inflection (see for instance Eide et al., 2010). Whether what may be termed a global journalism is emerging in relation to the issue of climate change is yet to be determined. A number of studies do show, however, that the level of attention is somewhat transnational (see Schäfer et al., 2011; and Shehata and Hopmann in this issue). Studies of attention cycles, however, have also shown that (international or domestic) political differences are somehow needed in order to attract journalistic and thus public attention (see for instance Brossard et al., 2004). Since climate change is often remote in time it somehow needs other concrete events—like extreme weather or summits—to be made the object of journalism.

As a social system with its own specific logic journalism transforms any issue from its starting points to its appearance as an event in the public debate. Such transformations into (political) events partly happen though various framings and cultural domestications; and the transformation from science to politics (and/or economics) through established journalistic norms may distort what is a broad scientific consensus into a "balanced" perspective, and thus contested issue, as Boykoff and Boykoff (2004, 2007) have pointed out, or into a "catastrophic" view, as Weingart et al. (2000) have shown. A related issue here is the lines separating objectivity from advocacy and the notion of communication for social change (see for instance Dilling and Moser, 2007 and Anita Howarth in this collection). Linked to this are studies of how journalism relates to the various vested interests involved in setting the agenda in relation to climate change, e.g. non-governmental organisations (NGOs) and different colours of think tanks, and the wider sociological settings in which this takes place (see for instance Hulme, 2009). The vast majority of empirical studies of the journalism on climate change have, however, focused on media content through various forms of frame-analysis and/or discourse analysis. Seen in relation to the consumption of journalism, there has been a growing interest in the formation of public opinion within the tradition of framing and media effects studies. What has been somewhat missing has been work within a wider audience studies tradition. More of such studies are, however, needed in order to gain a better understanding of some of the consequences of climate change communication, especially how this may or may not relate to social actions.

The proposal for a special issue of *Journalism Studies* on Environmental Journalism grew out of two conferences: "Responsibility Across Borders—Climate Change as Challenge for Intercultural Inquiry on Values" held at Aarhus University in Denmark in November 2009; and "Communicating Climate Change II—Global Goes Regional", a pre-conference of the ECREA 2010 in Hamburg in October 2010. Participants from these two conferences are among the contributors to this issue along with others who responded to an open call for papers. The complexities of the issue, some of which are touched upon above, and the ensuing interest from a number of divergent communication fields, have made the compiling of this special issue a challenging task. Yet, given that the overall goal was precisely to highlight the complexity of the issue, the aim was to select contributions from different disciplines using different theories, methodologies and, not least, focusing empirically on different parts of the world. In relation to that, a recurrent problem with the study of global phenomena is that certain parts of the world remain understudied. To some extent this has been the case with respect to the focus of the authors who responded to the paper call for this issue. Consequently, of the eight articles in this issue, only two focus on countries outside the Northern hemisphere (Bangladesh and Argentina), but these articles do provide valuable correctives to issues that are often taken for granted in the journalism studies of the North. Moreover, the research in this themed issue offers an additional and rich range of empirical studies in diverse national settings including Australia, France, Norway, Sweden, the United Kingdom and the United States.

With regard to focus and methodology, the majority of the articles in this issue are based on various types of content analysis of printed journalism, a leaning that mirrors the research field. Although print is only one element in the broader mediation of climate change, the issues raised in newspapers somehow relate to those of the wider field; also, a focus on print lends itself more easily to a comparison between different cultural contexts. There are also some articles, e.g. Bacon and Nash, which advocate looking beyond texts in order to understand the complexities surrounding the production of journalism on the environment and climate change. This is certainly a call that we hope will be taken up by researchers in the future. That said, however, one may also stress the wide variety of questions and topics that studies of content may bring up and address. The articles of this issue certainly attest to that.

The ordering of the articles in this issue has, however, not been based on geography but on the overall scope of the focus of each article. The first article, Stefan Cihan Aykut, Jean-Baptiste Comby and Hélène Guillemot's "Climate Change Controversies in French Mass Media 1990–2010", aims to understand the processes and actors, which interact in order to establish something as a public problem. Journalists are here seen as one part—the others being scientists, politicians and NGOs—of the shifting social configurations that frame, i.e. define and redefine, problems as (and in) public. Based on a theoretical mixture of the sociology of public problems, media sociology, science and technology studies, this article foregrounds the complex interplay of the processes in which journalism grapples with the issue of climate change. As such, it provides a very relevant background for the remaining articles.

The next two articles also deal with framing. The first of these, Adam Shehata and David Nicolas Hopmann's "Framing Climate Change: a study of US and Swedish press coverage of global warming", addresses the question of whether the framing of climate change is mainly influenced by the global scientific consensus or the discussions of domestic political elites. In order to explore this question, newspaper coverage from two countries in which political attitudes towards climate change differ, namely the United States and Sweden between 1997 and 2007, are analysed in relation to framing and the presence of various types of sources. The article concludes, contrary to other studies, that the coverage in these two countries was very similar in that controversies within the domestic political elites played a relatively minor role in the definition of the issue. The next article, María Teresa Mercado's "Media Representations of Climate Change in the Argentinean Press", deals with similar issues, and draws a somewhat related conclusion. Based on a content analysis focusing on (among other things) section, sources and themes of the coverage of the issue of climate change in two Argentinean elite newspapers in 2009–2010 it is argued that the issue in Argentina is dealt with primarily as a topic of international conflict related to what is described as an environmental debt.

The next two articles shift the focus towards issues of objectivity and advocacy. The first of these, Anita Howarth's "Participatory Politics, Environmental Journalism and Newspaper Campaigns", argues that traditional liberal notions of journalism and democracy stand in the way of understanding environmental journalism. In an era of disputed science and risks she argues, based on a case study of the coverage of genetically modified food in Britain, for an understanding of journalism related to notions of participation that are more appropriate in dealing with perceived risks and as such are to be seen as a form of empowering individuals outside the media. Jahnnabi Das' "Environmental Journalism in Bangladesh: active social agency" addresses similar concerns but in a wholly different context. Like Howarth's article, Das deals with issues that are highly relevant to the audience, namely river pollution in Bangladesh, and also shows that journalists here assume roles as active agents in the construction of the problem as well as in the formulation of its possible solutions.

The last three articles, similar to the lead article, shift the focus away from journalism itself and enlarge the setting in order to focus on some of the other major players that are active in setting the agenda on climate change, namely NGOs, business and think tanks. Roy Krøvel's "Setting the Agenda on Environmental News in Norway: NGOs and newspapers" aims to understand the developing relations between environmental NGOs and news journalism. He follows 17 Norwegian NGOs over a period of 10 years, 1999–2009, in order to evaluate their ability to set the agenda of the public debate on environmental issues in Norway. An important finding here is that a move towards more intangible risks has made more concrete, sensationalist NGO actions less efficient than a more long-term effort based on specific knowledge.

The final two articles deal with less overt attempts at agenda setting. Wendy Bacon and Chris Nash, in "Playing the Media Game: the relative (in)visibility of coal industry interests in media reporting of coal as a climate change issue in Australia", focus on how particular interests are, or are not, constructed journalistically. Through a Bourdieu-inspired field analysis important issues of visibility are highlighted in relation to the coverage of specific industry and climate change issues in Australia. The overall aim, apart from exposing specific Australian circumstances, is to argue that detailed studies of specific relations and levels of visibility are necessary in order to understand how various interests are an intricate part of constructing the public perception of climate change.

The final article also focuses on Australia. In "Talking Points Ammo: The use of neoliberal think tank fantasy themes to delegitimise scientific knowledge of climate change in Australian newspapers", Elaine McKewon follows and analyses the strategy of the conservative think tank, the Institute of Public Affairs (IPA), through a detailed discourse analysis of *The IPA Review*. Based on this, she looks at how the IPA channelled some of the established frames, claims, language and rhetoric into op-eds published in major Australian newspapers; and how conservative columnists took up and furthered some of the arguments within a frame of anti-elitist arguments that sought to discredit elite institutions of knowledge while arguing that these were disenfranchising those perceived outside such institutions.

"The unique complexity of climate change means", says Julie Doyle in her recent book *Mediating Climate Change*, that its "mediations span a wide range of institutions and representational practices, from positivist science and the environmental movement, to media representations, politics and popular culture" (2011, p. 2). Taken together the contributions to this issue portray how this complexity is given form within the institution of (printed) journalism in a number of divergent settings. At one level, the articles thus point towards the diversity of handling what may seem similar issues. But at another level, the articles also partly support Doyle's recent argument that the "original gap between scientific consensus on the anthropogenic causes of climate change and public and political acceptance of this issue" is no longer the main battleground. And the tension between balance and advocacy in relation to environmental journalism brought out by a couple of the articles point towards the new gap, which is "now increasingly related to the profound disconnect between political and public acceptance of climate change and the transformation of this knowledge into effective, urgent action to both mitigate and adapt" (Doyle, 2011, p. 2). What may be glimpsed in these articles is thus a certain shift in the role of journalists within a new global (media) landscape in which user communities with specific global concerns start to challenge the adherence of journalism to the maintenance of the nation state.

REFERENCES

BOYKOFF, MAXWELL T. and BOYKOFF, JULES M. (2004) "Balance of Bias: global warming and the US prestige press", *Global Environmental Change* 14, pp. 125–36.

BOYKOFF, MAXWELL T. and BOY, JULES M. (2007) "Climate Change and Journalistic Norms: a case study of U.S. mass-media coverage", *Geoforum* 38(6), pp. 1190–204.

BROSSARD, DOMINIQUE, SHANAHAN, JAMES and MCCOMAS, KATHERINE (2004) "Are Issue-cycles Culturally Constructed? A comparison of French and American coverage of global climate change", *Mass Communication and Society* 7(3), pp. 359–77.

DILLING, LISA and MOSER, SUSANNE C. (2007) *Creating a Climate for Change: communicating climate change and facilitating social change*, Cambridge: Cambridge University Press.

DOYLE, JULIE (2011) *Mediating Climate Change*, Farnham: Ashgate.

EIDE, ELIZABETH, KUNELIUS, RISTO and KUMPU, VILLE (Eds) (2010) *Global Climate—local journalisms. A transnational study of how media make sense of climate summits*, Global Journalism Research Series Vol. 3, Bochum/Freiburg: projekt verlag.

HULME, MIKE (2009) *Why We Disagree About Climate Change: understanding controversy, inaction and opportunity*, Cambridge: Cambridge University Press.

SCHÄFER, MIKE S., ANA, IVANOVA and ANDREAS, SCHMIDT (2011) "Globaler Klimawandel, transnationale Öffentlichkeit? Medienaufmerksamkeit für den Klimawandel in 23 Ländern", *Studies in Communication/Media* No. 1, pp. 131–48.

WEINGART, PETER, ENGELS, ANITA and PANSEGRAU, ELKE (2000) "Risks of Communication: discourses on climate change in science, politics and the mass media", *Public Understanding of Science* 9(3), pp. 261–83.

CLIMATE CHANGE CONTROVERSIES IN FRENCH MASS MEDIA 1990–2010

Stefan Cihan Aykut, Jean-Baptiste Comby, and **Hélène Guillemot**

This article offers an analysis of controversies surrounding the coverage of climate change in the French press. The theoretical framework for the analysis combines the sociology of public problems, media sociology, and science and technology studies. We present these controversies as an expression of a struggle over the ownership and framing of climate change as a public problem. Specific social groups are involved in this process of definition, framing, and agenda-setting. The success or failure of these groups in closing debates results in the construction of issues as either consensual matters-of-fact or controversial matters-of-concern. This framework allows us to distinguish between two phases in the career of this public problem, characterised by differences in ownership-configurations and the visibility of controversial points of view. We identify four relevant groups—scientists, politicians, journalists and non-governmental organisations—as well as certain social processes that help to explain changes in the attention that controversies received from the media. We conclude with the hypothesis of a third phase characterised by a relatively high degree of attention on controversies in French media.

Introduction

When environmental issues access the mass media, they often give rise to controversies (and publicity) concerning the accurate definition of problems, their urgency, and appropriate solutions. Such controversies are indicators of a competition between stakeholders over the framing of an issue. While some environmental issues are highly controversial, others remain or become more consensual. Differences can also be observed in the treatment of the same issue in different national settings. The aim here is to shed light on the treatment of such controversies in the French climate change debate. Specific social groups are, through processes of definition, framing, and agenda-setting, involved in the construction of environmental facts as public problems. Their success or failure in closing debates results in the construction of issues as either consensual matters-of-fact or controversial matters-of-concern (Latour, 2004). Since the definition of environmental issues is highly dependent on the products of scientific work, the classical focus of policy analysis on the mass media and political actors needs to be enlarged to scientists. We therefore propose to combine media studies and Science and Technology Studies (STS) within a sociology of public problems framework.

In order to investigate how the climate change issue became "publicised", our case study builds on firstly, a content analysis of three French publications over the period 1987–2007[1] and of television news on the two leading French channels (1997–2006),[2] and secondly, interviews with key actors from different domains who have contributed to the framing of the problem.[3]

The first section explains our theoretical approach and develops a first hypothesis: namely the visibility that controversies receive in the media varies between the different phases of the career of a public problem. Such a career is determined by changing social configurations among groups that define the issue, such as scientists, experts, non-governmental organisations (NGOs), state agencies, political groups, lobbyists, journalists, and in some cases parliament or the courts.[4] Using the concepts of "career" (Becker, 1963) and "configuration" (Elias, 1970) aims to strengthen both long-term and relational perspectives in order to overcome the frequent over-interpretation of the impact of specific causes such as single meteorological, political, cultural or economical events. Symbolic modifications in the framing of a problem, we suggest, depend on social transformations within and between stakeholders of a public problem in a given socio-historical context.

We use this approach in the following sections to analyse the public career of the climate change issue in France: in the second section, we distinguish two phases in the construction of climate change as a public problem. In the third, we identify four groups of actors involved in the framing process—scientists, politicians, journalists and NGO activists[5]—and four social processes that affected the configuration of these actors between phases one and two. In the fourth section, we look at the visibility of controversies in French mass media throughout the career of the public problem. We show that changes in the attention controversies received can be explained by the social processes affecting the actor configurations.

In the final section, we discuss the hypothesis of a third phase since 2009 in which the very fact that climate change has ascended to the top of political, scientific, and media agendas has contributed to making controversies more visible. We believe this increased salience has enlarged the range of stakeholders who wish to take part in defining the problem. This in turn intensified the framing contest. This is the reason why, while it is useful to consider controversies in studying the media construction of environmental issues, media studies can also help to improve the analysis of the dynamics of controversies.

Media Analysis, Public Problems, and Climate Change Controversies—Towards a Theoretical Framework

Media scholars have mainly addressed climate controversies by focusing on "climate sceptics" and their attempts to attack climate research and challenge the legitimacy of climate change as a social problem (Antilla, 2005; Lahsen, 2005; McCright and Dunlap, 2000). The success of these attacks has been invoked to explain political inaction with regard to the issue (McCright and Dunlap, 2003; Mooney, 2005; Oreskes and Conway, 2010). While such analyses offer valuable insights into the strategies and networks of dissenting voices, they are less suitable to explain why certain issues become controversial at a specific point in time and space. A larger framework that tries to answer this question has to include structural reasons for the success of actors in accessing the media and imposing their framing of an issue. Three such accounts/frameworks have been applied to the climate change debate.

Downs (1972) postulates the existence of an "issue attention cycle" for public problems and sees controversy as a *natural stage* in the historical deployment of an issue in the public sphere: the public's realisation of the costs and difficulties involved in the resolution of a problem inevitably blurs the formerly clear message of alarm, and the ensuing controversies lead to a decline of public interest. It has been argued that the American climate change debate follows such a pattern (Trumbo, 1996; Ungar, 1992). A key problem with this approach is its inability to account for cultural differences (Grundmann, 2007). Another account claims that the journalistic norm of "balanced reporting" leads to a "bias" in the media treatment of climate change, as equal media space is allocated to mainstream science and to the tiny minority of scientists challenging climate change (Boykoff and Boykoff, 2004). However, the explanatory variable of "journalistic norms" is clearly media-centred. In order to understand the full dynamics of how social problems gain access to and stay on the media agenda, we have to take into account journalists' sources and their strategies (Schlesinger, 1990). Following a third line of thought, controversies result from a dynamic of over-dramatisation and *backlash* through which the newsworthiness of the issue is maintained. In a first phase, actors from the journalistic and political fields gain public attention through the "over-dramatisation" of the issue, giving the most alarming scientific results high visibility. In a second phase, as other issues inevitably become more salient again, actors engaged in the debate realise the "risks of communication" they face (Weingart et al., 2000): scientists' credibility is on the line in public controversies, politicians fear voters' reactions against climate policies and journalists are concerned about the declining interest of their audiences. Such perceived risks produce a more discreet attitude especially among scientists. Some journalists, on the other hand, switch to another dramatic register to maintain public attention ("global warming is a fairy tale"), which allows "sceptics" easier access to the media.

Two comparative studies show that the early French climate discourse did not conform to any of these logics. The first analysis compares coverage of climate change from 1987 to 1997 in the French newspaper *Le Monde* and in the *New York Times* (Brossard et al., 2004). The authors argue that differences in media coverage are related to differences in culture, i.e. the opposition of American-type "investigative journalism" and French "opinion journalism", which favours story-telling and political events (see also Kuhn, 1995). As a result—and contrary to what an issue-attention account would suggest—climate change was treated in France *from the beginning* as a partisan issue. Mormont and Dasnoy (1995) confirm this finding of a more controversial and political debate in France in their case study comparing France, Germany, and Belgium. Their explanation does not rely solely on journalistic norms, but also includes sources and their strategies. The authors show how the different structures of the journalistic and scientific fields and the unequal development of environmental pressure groups led to different patterns of media representation of climate change.

Building on these findings, we seek to draw attention to the *evolutions* in the French media coverage of climate change that become visible when covering a more extended time-span than these studies. In order to do so, we propose a framework that takes into account dynamics in the journalistic field *and* among sources. Before we present this framework, we will use insights from the sociology of public problems and STS[6] to clarify two notions that are central to our study: the construction of public problems and controversies.

In his path-breaking account of the construction of drinking-driving as a public problem in the United States, Gusfield (1981) shows how groups of actors—so-called "owners" of the problem—compete for the framing of the issue, and ties the emergence of hegemonic framings to specific "ownership configurations". In a similar vein, Hilgartner and Bosk (1988) direct our attention to the competition between issues for the scarce resource of public attention. They suggest that issue arenas (the press, parliament, ministries, etc.) have natural "carrying capacities" that limit the number of issues under active consideration. In order to attract and maintain attention on an issue, promoters have to dramatise it and appeal to culturally entrenched codes and myths that make it intuitively intelligible. Seen through the lens of the sociology of public problems, Mormont and Dasnoy (1995) look at the ownership configuration of an issue, and at the success or failure of owners in putting the issue on the media agenda as well as their ability to impose their framing. Evolutions in such ownership configurations, we suggest, constitute the "career" (Becker, 1963) of a public problem.

Insights from research in STS help to deal with the complex issue of climate controversies. The term "controversy", given to debates or polemics on climate issues, is in itself deeply problematic. For climatologists, there is no controversy on climate change, because their opponents' arguments are not validated through the mechanisms of the scientific community; inversely, opponents claim that the absence of scientific controversy reflects the domination of an institution. As STS scholars have shown in a body of literature on "boundary work", debates on borders and on the legitimacy of actors and their findings are often part of the controversy (Gieryn, 1983; Jasanoff, 1987). Furthermore, climate change covers a large range of scientific domains and is a high-stakes political and economic issue. We decided not to restrict our analysis to controversies confined to the boundaries of disciplinary science, but instead to take into account much more varied, complex disputes. Such "socio-technical controversies" "overflow" into several arenas, combine different interests, involve different regimes of justification, and mix scientific, political, and ethical arguments (Pestre, 2003; Rip, 1986). We therefore concentrate on all kinds of "dissonant voices" that initiate and participate in disputes, debates, polemics, and controversies about the urgency, impacts, and social significance of climate change, the appropriateness of the methods of climate sciences, the efficacy of climate policies, as well as the merits of the precautionary principle and of what some perceive as a new "green" ideology. This allows us to draw a more fine-grained picture of the trajectory of the public problem, and of the evolution of resistance and critique it has encountered.

The Construction of Climate Change as a Public Problem in France

Although climate change was on the political agenda as early as in the late 1980s in France, its evolution into a full-fledged public problem only occurred in the early 2000s. We distinguish two periods in the construction of this public problem: the first from 1988 to 2001/2 and the second from 2002/3 to 2008.

Climate change became a political issue in France through international summits like the Toronto Conference on the Changing Atmosphere (1988), the Hague Conference on the Environment (1989) and the economic summit of the Group of Seven (1989) in Paris.[7] At about the same time, the success of green parties at municipal and European elections increased pressure on the political system.[8] As a consequence, the new government of then Prime Minister Michel Rocard (1988) appointed an inter-ministerial group to elaborate a French policy response (GIES, later MIES).[9] A conference on climate change was held in parliament, and the Ministry of Research organised a summit "Planet Earth" attended by President Mitterrand. However, no coherent domestic climate policy or reduction target was adopted and the political system's attention to the issue dropped sharply after the Rio conference in 1992. Moreover, the attention of major newspapers remained very modest throughout the 1990s. An analysis of three major French newspapers shows that major peaks in news attention resulted from international events like the 1992 Rio and 1997 Kyoto conferences (Figure 1). Indeed, climate change was not considered a major domestic problem, and climate policy was reduced mainly to its international component (Brossard et al., 2004). The organisation of expertise also contributed to the absence of a public debate on climate change: the reports of the GIES were confidential and their public impact limited, and a commissioned study by the National Academy of Science (1990) was too technical to gain widespread attention. We can thus conclude that despite intense activity in the political system at the very beginning of the period, climate change did not reach the top of the political or the media agenda before the 2000s (Aykut, forthcoming).

In the second phase, newspapers covered climate change increasingly through domestic meteorological phenomena (winter storms in 1999, a heat wave in 2003) and political events (national climate plans in 2000 and 2004). Peaks of media attention still coincided with international events like the Hague conference in 2000 and the release of the Third (2001) and Fourth (2007) Intergovernmental Panel on Climate Change (IPCC) Assessment Reports. However, coverage of the issue remained at a high level—and even continued to rise—between these events.[10] Climate change became a "structuring discourse" in the mass media[11] and began to be "institutionalised" in political

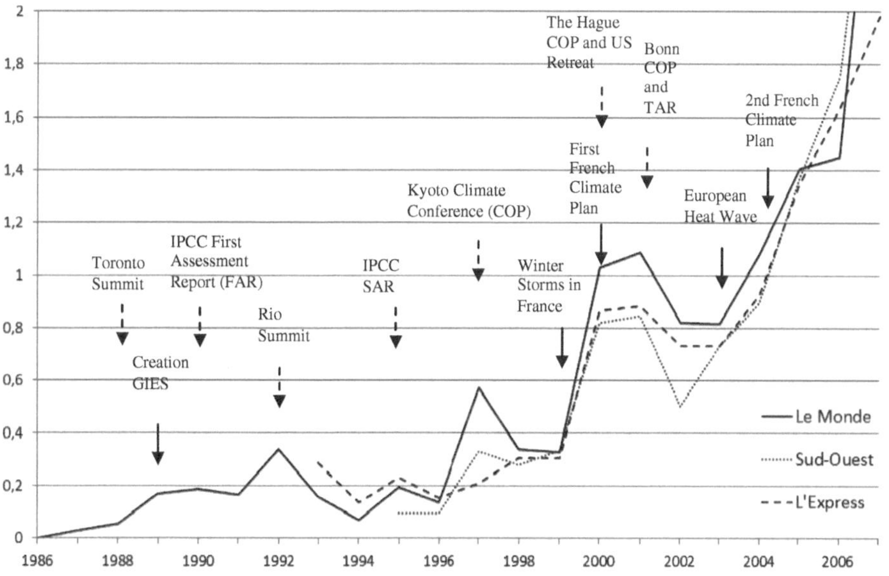

FIGURE 1

Climate change coverage in the French press. The vertical axis shows relative frequency of articles dealing with climate change (number of articles divided by editions). Major events covered in the press are indicated above the lines. Keyword search on Europress database: "changement(s) climatique(s)", "effet de serre", "réchauffement global", "réchauffement de la planète". For more details on the findings of the news analysis, see Aykut (2011)

programmes and organisations.[12] The creation of the National Observatory on the effects of climatic warming in France (ONERC) and the commissioning of a report on the effects of climate change on French geography by the two chambers of parliament (both 2001) illustrate this process of institutionalisation. In 2004, an Environmental Charter was added as an amendment to the French Constitution. Furthermore, the dominant framing of the climate issue changed. While in the first phase the issue had been framed as an issue of international relations, its construction as a public problem in the beginning of the 2000s was based on an "interpretative package" (Gamson and Modigliani, 1989) that consisted of two main assertions: human activities induce a supplementary greenhouse effect that leads to global warming, and everyone has to fight it primarily at an individual level, through lifestyle changes and energy savings (Comby, 2009).

Changes in the Ownership Configuration: Four Social Transformations

Two extreme weather events are often invoked to explain the increased media attention to the issue: a major storm in the winter of 1999 and a heat wave in 2003. Both cases were widely interpreted by the media as signs of global warming, even though scientists refused to publicly establish such clear links. We argue that these events certainly played a role, but that the increasing newsworthiness (Schlesinger, 1978) of climate change relied more on social shifts. In fact, the configuration of the owners of the public problem, their access to the media and to the political definition of the issue changed at the beginning of the 2000s, allowing for a different treatment of climate change in the media. We identify transformations in four fields—journalistic, scientific, state and NGO—that explain these changes.

By 2000, journalists dealing with environmental issues had gained a new status in newsrooms. Mormont and Dasnoy (1995, p. 54) show that the field of environmental journalism was poorly established in France in their period of analysis, leading to a more "political" framing. Comby (2009) argues that this changed as it became a full-fledged specialty by the beginning of the new millennium. In the process of the institutionalisation of environmental journalism, new journalists entered the field. Whereas journalists who traditionally dealt with environmental issues tended to interpret them in ways that were close to activists' points of view and promoted political frames, the new generation of journalists was driven by professional more than political motivations and produced information more closely adjusted to the commercial rules of the journalistic field. In order to increase the newsworthiness of the issue, "anecdotal" frames came to be favoured over "thematic" ones (Iyengar, 1991), and especially television news and popular media focused on "awareness raising" instead of providing more structural explanations. This means that the consequences of climate change were presented in terms of concrete effects on people's life-worlds, and options for mitigation were reduced to individual lifestyle changes. By doing so, environmental journalists contributed to shaping the issue as a matter of individual morality. Contrary to the "political" framing in the beginning of the 1990s, this new framing leaves less room for controversy.

Social and institutional changes can also be observed in the scientific field. Until the early 1990s, French climate research was weakly organised, climate change was not a major research issue in France, and most climatologists were not mobilised on this issue. This, as well as the lack of institutionalised structures for communication to the media, blurred the scientific message. Thus, the floor was left to popularisers and more political frames (Mormont and Dasnoy, 1995, pp. 54–5). Guillemot (2007, chaps 10 and 11) shows in detail how the French climate research community became more organised in the late 1990s. A key step in this process was the creation of the Institut Pierre Simon Laplace (IPSL) which facilitated the creation of a more comprehensive French climate model by coupling an atmospheric model and an ocean model.[13] Scientists formerly working on isolated topics began to work together, and different laboratory "cultures" were integrated in the modelling effort (Edwards, 2001). The completion of the model in 1997 and the growing institutionalisation of climate science in France changed the initial attitude of climatologists towards both the IPCC assessment exercise and the mass media. No French modelling results were included in the IPCC simulations for the first two assessment reports (1990 and 1995) because no coupled model was available at that time; even in the Third Report (2001), when French coupled models were completed, only four simulations were included, as scientists were reluctant to spend considerable amounts of time on activities they did not consider central to their research. The situation changed radically with the Fourth Report (2007): *two* different models provided simulations for *all* IPCC scenarios. Communication to the press also became progressively more professionalised (Tristani-Potteaux, 1997), and some climatologists began to invest time in communicating with or educating journalists. Scientists play a central role in legitimising environmental journalists in their newsrooms. Speaking about difficulties convincing her boss to run news stories about climate change at the end of the 1990s, a TV journalist said in one of our interviews in 2005: "we were struggling and I'm happy to hear that there now is a scientific consensus . . . I was always looking for the scientist who was willing to say 'yes, it is due to humans that . . .'". Another example of scientific mobilisation in defence of the newsworthiness of climate change is a "workshop" for editors-in-chief organised annually by Jean-Marc Jancovici, an expert on energy savings (Comby, 2009).

The State also played a central role in the public promotion of the problem. The activity of GIES, directed by a high-ranking public official (Yves Martin), is characteristic of the state's role in the first phase. This committee, which worked in a technical and confidential manner[14] (Roqueplo, 1993, pp. 204–5), regularly promoted the idea of a European carbon tax[15] to reduce CO_2 emissions (Aykut, forthcoming). The tax initiative finally failed and the policy window closed in 1997 with the adoption of the Kyoto protocol. In 2000/1 the political salience of climate change increased suddenly for two reasons: first, US President George W. Bush's decision not to ratify the Kyoto protocol polarised the issue and contributed to keeping it in media headlines for years. Second, France proposed its first comprehensive climate action programme. This new salience, combined with the unwillingness of the French government to push for a carbon tax after the failure of the European initiative, paved the way for a second actor, the national agency for energy efficiency and savings (ADEME), to frame the topic in accordance with its institutional objectives.[16] In 2004, it launched a communication campaign to promote individual energy savings, backing it with the claim that individuals are responsible for 50 per cent of French greenhouse gas emissions.[17] NGOs, journalists, the educational system, and some scientists took part in this "symbolic crusade" (Gusfield, 1981) that consisted in promoting lifestyle changes, understood as "greener" daily practices.

Finally, NGOs were engaged in multiple activities aimed at increasing the salience of the climate issue. Their activities included lobbying at different levels of government, the organisation of conferences and demonstrations, co-operation with schools and cultural institutions, and supplying expertise (Ollitrault, 2008). Up until the end of the 1990s French NGOs were poorly organised with regard to climate change and did not play a decisive role in promoting the issue. They gained major importance in the discussions before the Kyoto conference, as key environmental NGOs formed the French branch of the Climate Action Network in 1996. The subject was also taken up and popularised by a handful of charismatic leaders like, for example, Nicolas Hulot. Host of a very popular television show (*Ushuaïa* on TF1), he began his career as an environmental activist in the 1990s, when he founded the "Ushuaïa Foundation" (1990). His political influence grew in the 2000s: he is said to have generated the idea of the Environmental Charter (2004), and he convinced five candidates in the 2007 French Presidential Elections (including current President Sarkozy) to sign an "ecological pact". Hulot and Yann Arthus-Bertrand, a photographer influential in the French climate debate, also promoted the climate cause through films (*The Titanic Syndrome* and *Home*). Another feature of the activities of NGOs was their co-operation with state agencies. As an example, Hulot's "Earth challenge",[18] supported by ADEME, brought together dozens of celebrities and around 800,000 individuals on the Internet. As a result, NGO activities contributed to a "consensus mobilization" (Klandermans and Oegema, 1987) around the climate issue in the 2000s.

The conjunction of these social changes explains why climate change rose to the top of the hierarchy of public problems in France in the early 2000s. In the following section, we will show that the transition between different stages of the public problem's career also entailed changes in the visibility of different types of controversies. This should not come as a surprise, as social configurations determine both the place of a public problem in the newsroom *and* the means that different actors use in order to gain media visibility for their positions. We look at the general presentation of climate sciences in the media (consensual or uncertain), as well as the visibility and variety of controversies.

Controversies in the Career of the Public Problem

A qualitative analysis of the headings of the newspaper *Le Monde* between 1995 and 2006 provides a first insight into the differences between the two periods (Table 1). In the first phase, internal scientific controversies were translated in the mass media into a language of uncertainty. In accordance with the cautious stance of climatologists, articles stressed the complexity of climatology and of the "climate machine" and warned about drawing hasty conclusions. While they highlighted uncertainties, they nevertheless called for political action. From around 2003, these themes were no longer dominant and article headlines began to focus more on the drastic consequences of climate change. Responses to "climate sceptics" became commonplace and the tone of the debate more impassioned. Our interviews show that this evolution of journalists' treatment of the issue coincided with a changing attitude among French climatologists. A prominent climatologist, for example, says about this period that it had become more important to be persuasive about certainties and points of consensus than to stress limitations and uncertainties.

In order to get a clear picture of the evolution of controversies, we first categorise them according to the disciplines of contesting actors. Our typology distinguishes five types of actors, depending on their distance from the "core science" of climatology (or Working Group 1 of the IPCC). The *first* group includes climate scientists; actors from the

TABLE 1
Headlines of articles in *Le Monde* (translated samples)

Phase 1	Phase 2
Difficulties correspond to scientific uncertainties (13 March 1995)	Multiplication of extreme events worries climatologists (13 September 2003)
Climatic uncertainties for future cultures (4 April 1995)	"In 2100, temperatures will have increased by 1.4 to 6°C" (20 June 2004)
Scenarios about expected climate change still vague (23 September 1995)	The Earth, sick of mankind (20 June 2004)
The computer is an expert, but not a prophet (18 February 1997)	On Realclimate.org, specialists track scientific fraud and errors (2 January 2004)
"It's too early to draw conclusions" (28 May 1997)	Warming of more than 6°C is not excluded (30 January 2005)
The terrestrial "climate machine" challenges science (5 January 2000)	Sea-level rise could be higher and quicker than predicted (25 March 2006)
Uncertainty undermines predictions (3 October 2001)	Cyclones: more frequent or more violent? (6 September 2006)
Concerning sea-level rise, a reliable method for calculation is yet to be found (8 November 2001)	Europe is experiencing its hottest autumn in five centuries (8 December 2006)

second group are scientists from disciplines involved in impact research (ecology, geography, etc.); the *third* group consists of people within disciplines concerned with mitigation (mainly economists); actors from the *fourth* group are scientists and engineers from other areas (mathematics, physics, geophysics, etc.); and actors belonging to the *fifth* group are intellectuals or social scientists not professionally concerned with climate change (philosophers, writers, journalists, etc.).

In the *first* phase, different kinds of dissonant voices expressed themselves in the press, and even though they never dominated the media coverage of the issue, they were by and large tolerated. Throughout the two periods, they mainly consist of public figures from the fourth group who share a specific profile: Haroun Tazieff in the first period, as well as Claude Allègre and Vincent Courtillot in the second, are at once highly recognised as scientists in domains not related to climate science—all three come from the geosciences—and former ministers or high public officials. They thus combine scientific, political, and media capital, have influential networks in the National Science Academy, and are well known to both scientific and political journalists. Yves Lenoir, an engineer and activist against the French nuclear programme, is another important actor in the first phase (also group four). He argued against the urgency of the climate issue in the press, and published a book promising to unveil "the truth about the greenhouse effect" (Lenoir, 1992). Publishing books is a strategy often used by French contesting actors to overcome the problem of scientific legitimacy they face, as they generally lack publications in academic journals.

It is important to understand the intellectual environment in which the debate on climate change took place. In France in the 1990s, the growing political focus on ecology provoked a strong resistance from a part of the political, intellectual and scientific elite against what they perceived as a new "ideology" or "new ecological order" (Ferry, 1992). This counter-movement, formed around resistance to the Rio conference and the introduction of the precautionary principle in France and Europe, culminated in 1992 in the Heidelberg appeal,[19] which was initiated by the French physician Michel Salomon and signed by hundreds of high-profile scientists (including 72 Nobel Prize winners). This had a significant impact on the French press. Contrary to the US context (McCright and Dunlap, 2003; Oreskes and Conway, 2010), contesting actors in France have usually had little contact with industry and political cleavages have not been played out in the climate field. Contesting actors come instead from the scientific, administrative, and political establishment (often the French Socialist Party) and defend a certain vision of science, as well as of the relationship between science and the state. In the Heidelberg appeal, they presented the values of state-led progress as being endangered by "irrational environmentalism", to which they opposed "scientific environmentalism". While the appeal was not explicitly aimed at climate change, it spoke against nascent preoccupations with/about the global environment and the political initiatives launched to respond to the ecological crisis. The preparatory group, called "Reflection Committee for a Democratic Environmentalism", was made up of scientists from different disciplines and "intellectuals against the green lobby" (*Le Monde*, 19 June 1992). Some of them, such as Haroun Tazieff, Luc Ferry, Michel Onfray, and Pascal Bruckner, later participated in attacks against climate science or policy, or in controversies about its consequences (Hervé Le Bras).

Although "contesting actors" stayed active, the overall visibility of controversies changed in the *second phase*. Scientists' presence in the press increased, and less public attention was given to dissonant voices. Actors not sharing the consensus about the reality and urgency of anthropogenic climate change have since then been labelled "sceptics"[20] and owners of the problem (essentially environmental journalists and scientists) restrict their access to public discourse. This became explicit in an episode in 2006, as journalists, together with climate scientists, condemned (in a letter to the Science Academy, in articles and interviews, etc.) a sceptical article by Claude Allègre published in *L'Express* (Allègre, 2006).

A content-analysis of *Le Monde* articles (Figure 2) provides a more fine-grained picture of the evolution of controversies in the transition between phases one and two. The analysis shows a shift in the topic of controversies in this newspaper around the turn of the millennium: while the first phase was characterised by sporadic but persistent attacks on the hypothesis of human-induced climate change and the "new green ideology" expressed in international climate negotiations, such voices received little media attention in the second phase. Interestingly, the fact that climate change was already considered a reality and an issue for public policy opened up some space for controversies of a new kind on the margins of the dominant frame.

A qualitative analysis shows that controversies on climate change impacts tended to be "soft", meaning they were presented as "normal" scientific controversies without the polemic rhetoric prominent in the first phase. This was also visible in TV news, as some reports mentioned scientific uncertainties in relation to the consequences of climate change while more generalising statements about the uncertainties of climate science and modelling disappeared. Likewise, economic controversies obtained public visibility without questioning key elements of the dominant frame such as the reality and urgency of climate change. Another new development in the second phase was the explicitly "scientific" challenge by French Academy of Science member Vincent Courtillot. Concerned with the role of the sun and of geomagnetic fields, he positioned himself as a member of the scientific community engaged in an academic dispute. This positioning, as well as leading climate scientists' refusal to endorse it, can be understood as a form of

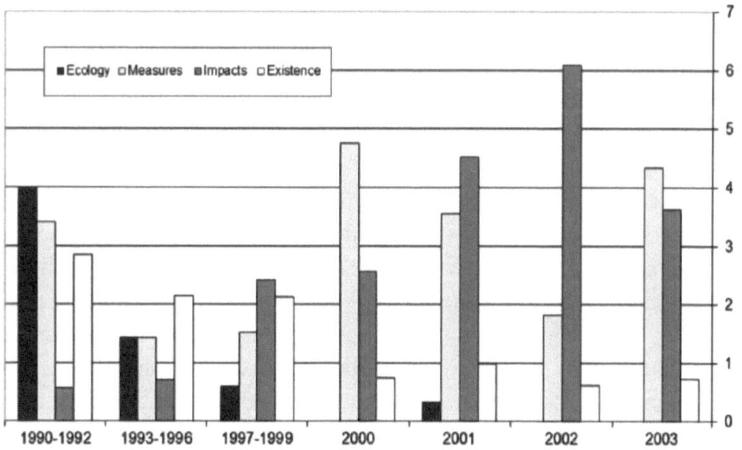

FIGURE 2
Controversies in *Le Monde* and transition between phases one and two. Our corpus includes 1531 *Le Monde* articles dealing with climate change (keywords: "changement(s) climatique(s)", "réchauffement climatique"/"réchauffement de la planète", "effet de serre"). From this sample, we extracted: (1) articles reporting about a socio-technical controversy; (2) articles where an actor takes a controversial position on a subject and responds to that article. The graph shows controversial articles in relation to the total number of articles on climate change. The years 1990–1992, 1993–1996 and 1997–1999 were taken together for purposes of statistical significance. The vertical axis shows percentages. Four categories of controversies were distinguished: critics of political ecology and exaggerated reporting on climate risks (a framing in which mostly group five actors are involved in France), polemics on measures and economics (group three), controversies on climate change impacts and consequences (group two), challenges of the existence of climate change (mostly group four). The first result is that throughout the period of observation, controversies represented a small percentage of the whole sample of articles (maximum 10 per cent). Controversies about the existence of anthropogenic climate change and critics of environmentalism were relatively common at the beginning of the 1990s and diminish sharply moving into the 2000s. On the contrary, controversies over impacts increased steadily throughout the 1990s, reaching relatively stable visibility in the 2000s. Controversies over economic measures were present throughout the observed periods, reflecting political discussions about eco-taxes (1992, 2000), tradable permits (1997) and national policies (2000, 2003). Their cyclical pattern suggests some independence with regard to the rest of the climate change debate

"boundary work". The fact that legitimacy in this particular scientific field is perceived as an important asset in public discourse illustrates climate scientists' central position in the ownership configuration.

However, not everything changed between the two phases, as two general observations about specificities of the French climate discourse demonstrate: first, international controversies did not play any role in the period of observation, even though they were very salient in the US discourse in the 1990s (Boykoff and Boykoff, 2004; McCright and Dunlap, 2000), and had an impact elsewhere, for instance in Germany (Weingart et al., 2000). The second general observation is that group one actors (climatologists) did not initiate public controversies. Instead, they presented a unified front against dissonant voices.

A More Controversial Third Phase?

By 2009 a third phase may have begun in which controversies of all kinds obtain greater public visibility. Beginning in autumn 2009, two international controversies received substantial media coverage: the so-called "Climategate" resulting from the publication of leaked emails from the University of East Anglia, and the revelation of errors in the fourth IPCC assessment report concerning the melting of Himalayan glaciers. Domestic controversies were also given significant play in the media. These included debates after the publication of a book called *The Climate Imposture* by Claude Allègre, and the mobilisation of some geographers, culminating in a conference with the provocative title "No, Heaven Won't Fall on Our Heads" (French Association of Geographers, 16 September 2010). Our theoretical framework allows us to explain these recent developments in social terms and through a historically informed account of the French climate debate. To do so, we will first look at both changes and continuities between this hypothetical third phase and the preceding ones.

A noteworthy change in French climate discourse since 2009 is that new contesting actors and new themes gain access to the media. This concerns most notably scientists who contest claims about the consequences of climate change (group two) and intellectuals (group five). Furthermore, dissonant voices from group four (such as Courtillot) have more frequent access to a wider range of media spaces—newspapers, magazines, and talk shows. Accordingly, controversies have diversified, ranging from climate change impacts to the science–policy relationship, expertise processes, climate sciences, computer modelling, environmental ideology, and appropriate measures. Discourses often mix scientific, political, economic, and epistemic arguments. This allows some of these actors, albeit speaking as individuals, to *condense* several types of disputes or criticisms, thus gaining a wider audience and support. However, from a historical point of view, neither the subjects nor the actors of phase three controversies are new. Actors from groups two and five expressed their critiques *mezzo voce* before phase three, in confidential arenas such as websites and scientific magazines. The new configuration now allows them to be heard by a wider audience.

How can the increased visibility of dissonant voices be explained? Observers frequently invoke the "Climategate" controversy and the failure of the climate conference in Copenhagen in December 2009. We think, however, that changes in the underlying social conditions play a central role. Our hypothesis is that the very fact that the public problem ascended to the top of media and political agendas contributed to its more controversial framing. Climate change reconfigures many domains—ranging from scientific disciplines to the organisation of newsrooms and priorities in public policy—and new actors come to be directly concerned with the issue. Additionally, as a high-profile issue, climate change yields media, scientific and political capital. Events like Copenhagen and "Climategate" have merely accelerated an on-going process, as new actors with other priorities, working methods and world-views have emerged and challenge the existing ownership structure. In what follows, we will show the importance of such social evolutions in the journalistic and the scientific fields.

Scientific and environmental journalists are no longer the only ones who shape the way climate change is treated by the media. The climate issue has increasingly become a topic of interest for journalists specialised in political, social, and economic domains. While this extension of concerned journalists enlarged the carrying capacity of the journalistic field and explains the constancy of high levels of media attention in the second part of the

2000s (Figure 1), the new journalists also have other priorities and are more open to dissonant voices. They worry less about the scientific legitimacy of alternative sources than about their political and social influence, and give more value to the originality of a story.

On the scientists' side, a similar broadening of visible actors can be observed. As the tools and methods of climate sciences become important issues in the debate, climatologists face critics from scientists belonging to other disciplines concerned by climate change (group two). Climate models impinge on territories where other disciplines claim legitimacy, leading to competition between different epistemic cultures. Geographers, for example, generally rely on bottom-up approaches to analyse regional climates, whereas climate modellers tend to "downscale" from global circulation models. In the French context, this led to controversies over how to appropriately characterise the consequences of climate change.[21]

It should be noted that even in phase three, controversies reach mainly intellectual parts of the media field such as newspapers and debate shows, while they remain relatively absent from media platforms with larger audiences. Indeed, controversies have had unequal access to different media spaces throughout the three phases. Thus, in 663 reports on climate change in the two main French channels' evening news between 1997 and 2006, none of eight interviewed people questioned the scientific consensus, and between July 2009 and July 2010, only two of 117 news reports dedicated to climate change mentioned controversies. Furthermore, framings that promote the climate issue as crucial still prevail and dissonant voices continue to provoke strong collective reactions from the owners of the problem: as a response to Claude Allègre's book, over 600 climate scientists signed a petition asking the Ministry of Research to publicly express its confidence in their work.

The main results of our analysis are summarised in Figure 3, including the hypothesis of a more controversial phase three.

Conclusion

In this article, we set out to analyse controversies in climate change coverage through a theoretical framework combining the sociology of public problems, media sociology, and science and technology studies. This led us to view them as the expression of struggles over the ownership and the framing of climate change as a public problem.

A first result is that media coverage of climate change in France did not undergo the classical issue-attention cycle. Instead, media attention steadily increased and recently reached unprecedented heights. This can be explained by the extension of the carrying capacity for the issue in the journalistic field, as climate change became a subject for new journalistic specialties.

Secondly, we identify two phases in the public career of climate change. In a nutshell, the first phase is characterised by a poorly established field of environmental journalism, a fragmented scientific field, with scientists reluctant to work with the IPCC and to talk to the media, and a political system defining climate change as a problem of international relations, with little salience in domestic politics. Around 2000–2003, the scientific field became more unified and the IPCC was recognised as a major, uniting actor, leading scientists to become more deeply involved in public discourse. At about the same time, the institutionalisation of environmental journalism in most national media and of

	Phase one 1990–2001/2	Phase two 2002/3–2009	Phase three –2009
Main characteristics	• CC in the process of becoming a public problem • Ownership not settled	• High-profile public problem • Interpretative package • Highly structured ownership	• New actors, diversification of topics • Ownership of phase two challenged
Types/forms of controversies	• Reality and urgency of CC contested • Criticism of political ecology • Strong focus on uncertainties • Controversies tolerated	• Very few controversies on reality and urgency of CC, strongly disqualified • Uncertainties and "soft" controversies tolerated	• Reality and urgency of CC contested • Criticism of political ecology, climate sciences, IPCC • Controversies on impacts and measures
Arenas where controversies appear	• Newspapers • Some talk shows • Some books	• Confined to analyses and debate columns in news magazines • Some blogs • Talk shows • Some books	• Analyses and debate columns in print news • Twice on TV news • Numerous blogs • High presence in radio and TV talk shows • Numerous books
Visibility of contesting actors			
Group one	None	None	None
Group two	Low	Low	Some
Group three	Some	Some	Some
Group four	High	Some	High
Group five	Very low	None	Some

FIGURE 3

Summary table on the three phases. CC, climate change

climate policy in political and administrative arenas united actors from these fields and altered their ways of engaging with the issue. With this change in the ownership of the public problem, the conditions of access of critical voices were modified and scientific controversies and alternative frames were marginalised.

Finally, we suggest that we may have recently entered a third phase. The issue has become a high-profile public problem with exceptional media attention. However, the increase in the carrying capacity of public arenas comes at a price: as new actors enter the field, the dominant ownership configuration is challenged and controversies (re-)emerge.

Further research should broaden both the range of potential owners, including economic actors in particular, and the range of media platforms considered. Indeed, controversies are unequally visible in different public spaces, indicating different ownership configurations. Differences between traditional media and the Internet, debate shows and television news, or opinion columns and news articles offer rich empirical material for a topology of the social space in which controversies develop and gain visibility.

NOTES

1. Three types of newspapers were analysed: a national newspaper, *Le Monde* (1987–2007), a regional newspaper, *Sud Ouest* (1994–2007), and a weekly magazine, *L'Express* (1993–2007).
2. We analysed 663 reports from TF1 and France2 evening news between 1997 and 2006 (source: database of the National Audiovisual Institute) and found very little evidence for controversies. We therefore do not present a detailed discussion of controversies in television news. However, the fact that controversies are more or less visible depending on the media is in itself an important finding.
3. Our respective PhD theses include a total of 112 interviews of journalists, climate scientists, activists, and state officials. Most media analyses of climate change are based on content analyses. Combining both content analysis and interviews aims at investigating the social conditions of news productions.
4. This hypothesis was developed at an international conference on climate change controversies in mass media organised by the authors on 20 and 21 September 2010 in Paris, France. We thank the Groupement d'intérêt scientifique Climat Environnement Société and the Institut des sciences de la communication du CNRS for their financial and logistical support.
5. We consider these four actors as central. Industry, for example, did not actively engage in the framing contest in France, leaving the initiative to others (Comby, 2009).
6. A previous analysis of the French controversy over genetically modified organisms has demonstrated the fruitfulness of combining these two strands of literature (Bonneuil et al., 2008).
7. A notable difference between climate change and other environmental problems is that the alarm was sounded in international scientific circles. The problem is therefore characterised not by a "rise in generality", but by a phenomenon of "downscaling" to national contexts.
8. The French Greens obtained 9 and 10.6 per cent in two elections in 1989.
9. Founded in 1989, the "Inter-ministerial Group on the Greenhouse Effect" was transformed into a permanent "mission" in 1990.
10. On the evening news of the two main TV channels between 1997 and 2006, 60 per cent of the reports dealing with climate change were concentrated in the period after 2003.
11. After 2000, newspaper coverage of global warming averaged one or more articles per day.
12. Marteen Hajer (1993, pp. 45–6) draws this distinction between discourse structuration and institutionalisation.
13. Supplementary modules were added later.
14. Reports on the topic were published and written in terms not easily accessible for the lay public.
15. By 1990 nuclear power generation had come to represent over 70 per cent of total electricity production in France, making the country one of the least vulnerable to a carbon tax in the European Community.

16. ADEME is the result of a merger (1992) of three formerly independent agencies including the French agency for energy efficiency created after the oil crisis in the 1970s. After a decade of low oil prices and nuclear over-capacity, climate change gave renewed impetus to the topic of energy savings.

17. This is a statistical trick, as the 50 per cent are mainly due to emissions from transport and construction. It is misleading to identify them as "individual" emissions.

18. See http://www.defipourlaterre.org/.

19. See http://www.his.com/ ~sepp/policy%20declarations/heidelberg_appeal.html.

20. The denomination "sceptic" was imported from the American debate, but other terms like "denialism" or "contrarian scientists" never imposed themselves in the French discourse.

21. Again, this is not new: geographer-climatologist Marcel Leroux became a prominent figure amongst sceptical publics in the first phase through the publication of a number of articles. However, he scarcely received any attention in mass media.

REFERENCES

ALLÈGRE, CLAUDE (2006) "Le droit au doute scientifique", *Le Monde*, 27 October.

ANTILLA, LIISA (2005) "Climate of Scepticism: US newspaper coverage of the science of climate change", *Global Environmental Change* 15(4), pp. 45–72.

AYKUT, STEFAN C. (forthcoming) "La construction du changement climatique comme problème public. Débat médiatique et expertise climatique en France et en Allemagne", in: Selma Bellal, Benjamin Denis and Denis Duez (Eds), *Savoir experts, savoirs profanes: vers de nouveaux modes de construction de l'actualité politique?*, Bruxelles: Publications des Facultés universitaires Saint-Louis.

BECKER, HOWARD (1963) *Outsiders. Studies in the sociology of deviance*, New York: The Free Press.

BONNEUIL, CHRISTOPHE, JOLY, PIERRE-BENOîT and MARRIS, CLAIRE (2008) "Disentrenching Experiment. The construction of GM-crop field trials as a social problem in France", *Science, Technology and Human Values* 33(2), pp. 201–29.

BOYKOFF, MAXWELL T. and BOYKOFF, JULES M. (2004) "Balance as Bias: global warming and the US prestige press", *Global Environmental Change* 14, pp. 125–36.

BROSSARD, DOMINIQUE, SHANAHAN, JAMES and MCCOMAS, KATHERINE (2004) "Are Issue-cycles Culturally Constructed? A comparison of French and American coverage of global climate change", *Mass Communication and Society* 7(3), pp. 359–77.

COMBY, JEAN-BAPTISTE (2009) "Quand l'environnement devient médiatique. Conditions et effets de l'institutionnalisation d'une spécialité journalistique", *Réseaux* 157–8, pp. 157–90.

DOWNS, ANTHONY (1972) "Up and Down with Ecology. The 'issue-attention cycle'", *Public Interest* 28, pp. 38–50.

EDWARDS, PAUL N. (2001) "Representing the Global Atmosphere: computer models, data, and knowledge about climate change", in: Clark A. Miller and Paul N. Edwards (Eds), *Changing the Atmosphere: expert knowledge and environmental governance*, Cambridge, MA: The MIT Press.

ELIAS, NORBERT (1970) *Was ist Soziologie?*, München: Juventa.

FERRY, LUC (1992) *Le nouvel ordre écologique. L'arbre, l'animal, et l'homme*, Paris.

GAMSON, WILLIAM and MODIGLIANI, ANDRE (1989) "Media Discourse and Public Opinion on Nuclear Power: a constructionist approach", *American Journal of Sociology* 95(1), pp. 1–37.

GIERYN, THOMAS F. (1983) "Boundary-work and the Demarcation of Science from Non-science: strains and interests in professional ideologies of scientists", *American Sociological Review* 48(6), pp. 781–95.

GRUNDMANN, REINER (2007) "Climate Change and Knowledge Politics", *Environmental Politics* 16(3), pp. 414–32.

GUILLEMOT, HÉLÈNE (2007) "La modélisation du climat en France des années 1970 aux années 2000. Histoire, pratiques, enjeux politiques", Thèse de doctorat en histoire des sciences, École des Hautes Études en Sciences Sociales, Paris.

GUSFIELD, JOSEPH (1981) *The Culture of Public Problems. Drinking-driving and the symbolic order*, Chicago: Chicago University Press.

HAJER, MARTEEN (1993) "Discourse Coalitions and the Institutionalization of Practice: the case of acid rain in Britain", in: Frank Fischer and John Forester (Eds), *The Argumentative Turn in Policy Analysis and Planning*, Durham, NC and London: Duke University Press.

HILGARTNER, STEPHEN and BOSK, CHARLES L. (1988) "The Rise and Fall of Social Problems: a public arenas model", *American Journal of Sociology* 94, pp. 53–78.

IYENGAR, SHANTO (1991) *Is Anyone Responsible? How television frames political issues*, Chicago: Chicago University Press.

JASANOFF, SHEILA (1987) "Contested Boundaries in Policy-relevant Science", *Social Studies of Science* 17(2), pp. 195–230.

KLANDERMANS, BERT and OEGEMA, DIRK (1987) "Potentials, Networks, Motivations, and Barriers: steps towards participation in social movements", *American Sociological Review* 4, pp. 519–31.

KUHN, RAYMOND (1995) *The Media in France*, New York: Routledge.

LAHSEN, MYANNA (2005) "Technocracy, Democracy, and U.S. Climate Politics: the need for demarcations", *Science, Technology and Human Values* 30(1), pp. 137–69.

LATOUR, BRUNO (2004) "Why Has Critique Run Out of Steam? From matters of fact to matters of concern", *Critical Inquiry* 30(2), pp. 225–48.

LENOIR, YVES (1992) *La vérité sur l'effet de serre*, Paris: La Découverte.

MCCRIGHT, AARON M. and DUNLAP, RILEY E. (2000) "Challenging Global Warming as a Social Problem: an analysis of the conservative movement's counter-claims", *Social Problems* 47(4), pp. 227–48.

MCCRIGHT, AARON M. and DUNLAP, RILEY E. (2003) "Defeating Kyoto: the conservative movement's impact on US climate policy", *Social Problems* 50(3), pp. 348–73.

MOONEY, CHRIS (2005) *The Republican War on Science*, New York: Basic Books.

MORMONT, MARC and DASNOY, CHRISTINE (1995) "Source Strategies and the Mediatization of Climate Change", *Media, Culture and Society* 17, pp. 49–64.

OLLITRAULT, SYLVIE (2008) *Militer pour la planète: sociologie des écologistes*, Rennes: PUR, Res Publica.

ORESKES, NAOMI and CONWAY, ERIK M. (2010) *Merchands of Doubt: how a handful of scientists obscured the truth on issues from tobacco smoke to global warming*, London: Bloomsbury Press.

PESTRE, DOMINIQUE (2003) "Regimes of Knowledge Production in Society: towards a more political and social reading", *Minerva* 41(3), pp. 245–61.

RIP, ARIE (1986) "Controversies as Informal Technology Assessment", *Knowledge* 8(2), pp. 349–71.

ROQUEPLO, PHILIPPE (1993) *Climats sous Surveillance: limites et conditions de l' expertise scientifique*, Paris: Economica.

SCHLESINGER, PHILIP (1978) *Putting Reality Together*, London: Methuen.

SCHLESINGER, PHILIP (1990) "Rethinking the Sociology of Journalism: source strategies and the limits of media-centrism", in: Marjorie Ferguson (Ed.), *Public Communication: the new imperatives*, London: Sage.

TRISTANI-POTTEAUX, FRANÇOISE (1997) *Les journalistes scientifiques, médiateurs des savoirs*, Paris: Economica.

TRUMBO, CRAIG W. (1996) "Constructing Climate Change: claims and frames in US news coverage of an environmental issue", *Public Understanding of Science* 5(3), pp. 269–83.

UNGAR, SHELDON (1992) "The rise and (relative) decline of global warming as a social problem", *The Sociological Quaterly* 33(4), pp. 483–501.

WEINGART, PETER, ENGELS, ANITA and PANSEGRAU, PETRA (2000) "Risks of Communication: discourses on climate change in science, politics and the mass media", *Public Understanding of Science* 9(3), pp. 261–83.

FRAMING CLIMATE CHANGE

A study of US and Swedish press coverage of global warming

Adam Shehata and **David Nicolas Hopmann**

This comparative study investigates news coverage of climate change in the United States and Sweden. The main research question concerns the extent to which news coverage of climate change is influenced by domestic political elite discussion or the scientific consensus surrounding the issue. While there has been a widespread consensus in Sweden that climate change is (partly) caused by human activity and that there is an unquestionable need to take countermeasures, there has been substantial debate about the causes and the necessity of political action in the United States. Based on an extensive content analysis of 1785 articles over a 10-year period, as well as an intensive analysis of news coverage of the Kyoto and Bali summits, results show that media coverage is strikingly similar in these two countries, indicating a weak influence of national political elites on how climate change is framed in news coverage.

Introduction

Few public issues have gained such significant attention internationally as climate change. During the last decade climate change has moved from being a question of concern for the scientific community to achieving the status of a major political problem in need of urgent action (Giddens, 2009). The substantial increase in public attention surrounding global warming has led to intense debates regarding the role of the news media in reporting on climate change as a political issue (Antilla, 2005; Boyce and Lewis, 2009; Boykoff and Boykoff, 2004). Despite sometimes widely opposing opinions concerning the character of climate change coverage in the media, few analysts doubt that the news media have a crucial role for citizens' knowledge, perceptions and attitudes towards global warming. In short, what and how the media report on unobtrusive issues such as global warming has consequences for public perceptions and opinions (McCombs, 2004; Scheufele, 1999).

This study focuses on the issue of climate change in order to investigate two competing claims regarding the agenda- and frame-building processes that shape news coverage in western societies. On the one hand, some argue that news coverage of global issues and events is strikingly similar across countries due to the globalization of information flows through news agencies, as well as shared professional norms and values among journalists in western democracies (Reigert, 1998; Shoemaker and Reese, 1996). Others argue, on the other hand, that national news coverage of global issues is heavily influenced by domestic political circumstances in general, and indexed to the range of viewpoints within the domestic political elite in particular (Bennett, 1990; Entman, 2004).

We argue that comparing American and Swedish news coverage of global warming casts an important light on how agenda- and frame-building processes work in mediated democracies. While the scientific evidence of global warming and its causes has been widely available to both American and Swedish journalists since the 1990s, the political climate surrounding the issue has differed substantially between the two countries. While a broad political consensus among government and opposition parties has characterized Swedish climate change debate for a long time, this has not been the case in the United States. Here the level of elite conflict in the climate change debate has been substantially more intense, with influential political actors—such as President George W. Bush—actively challenging and questioning the problem as defined by the Intergovernmental Panel on Climate Change (IPCC) and others.

This study investigates to what extent such political differences are reflected in the news coverage of climate change in two Swedish and two American national newspapers (*Dagens Nyheter*, *Svenska Dagbladet*, *New York Times* and *Washington Post*). These outlets were selected because they are commonly considered as elite quality papers that share similar functions within each country. Based on a content analysis of climate change coverage over a 10-year period, as well as an intensive content analysis of the Kyoto (1997) and Bali (2007) summits on climate change, we investigate patterns of source use and the presence of several issue-specific frames related to the debate on global warming. The results are discussed in light of the literature on agenda building, frame building and international news flows.

International News Flows and the Domestication of Political Issues

Past research has shown that news coverage of global issues is influenced by a variety of factors at different analytical levels (Shoemaker and Reese, 1996). While some factors work to homogenize news content across national borders, other forces operate in the opposite direction. Climate change, as any other global issue, is subject to filtering processes at various gatekeeping levels within the international information system. These filters influence the amount of media attention devoted to global warming (*agenda building*) as well as how global warming is framed in the news media (*frame building*) (Entman, 1993; Gandy, 1982; Scheufele, 1999). At the very heart of this debate is the question regarding the relative influence of domestic elite actors, international organizations and inter-media institutionalization of international news on media content (Bennett, 1990; Rai and Cottle, 2007; Sparrow, 1999).

Causes for Cross-national Variation: Why Climate Change Coverage Might Differ

In this article we investigate the extent to which news coverage of climate change is pegged to, and shaped by, activities taking place within major domestic official institutions. The idea that news coverage closely follows the lines, debates and actions of domestic official institutions comes from research into both foreign and domestic policy (Bennett and Paletz, 1994; Hall et al., 1978; Wolfsfeld and Sheafer, 2006). This tendency of journalists to narrow public debate of political issues to key actors within official institutions has most succinctly been captured by Bennett's indexing hypothesis. In its original formulation the indexing hypothesis states that "mass media news professionals, from the boardroom to the beat, tend to 'index' the range of voices and viewpoints in both news and editorials according to the range of views expressed in mainstream government debate about a given topic" (Bennett, 1990, p. 106). More than anything else, the indexing hypothesis states that news coverage of political issues is driven by, and dependent on, elite political actors, while unofficial actors tend to have a secondary role at best. Their access to public debate in the media is constrained and conditioned by a variety of factors. Using the vocabulary proposed by Molotch and Lester (1974), elite political actors tend to have "habitual access" to the news media because of their inherent news value, while unofficial actors have to rely on "disruptive access". The indexing practice, it is commonly argued, is the combined result of a journalistic dependence on official actors, norms of objectivity as well as news values that favor actors in influential positions (Bennett et al., 2007; Hall et al., 1978; Schoenbach et al., 2001).

The empirical evidence for the indexing hypothesis is mixed. Focusing on foreign policy crisis events between 1945 and 1991, Zaller and Chiu (1996) found a strong correlation between press slant and congressional opinion in the United States, supporting Bennett's notion of indexing. Mermin (1999) found similar evidence of indexing in an analysis of news coverage of the American invasion of Panama, Grenada and the build-up to the 1991 Gulf war. Others have argued, on the other hand, that while journalists tend to follow and report the lines of domestic elite debate closely, they are not doing it slavishly. Althaus' study (2003) of American coverage of the Gulf war showed, in contrast to previous studies, that "television news did not merely shadow the debate occurring among US officials. Journalists frequently presented competing perspectives and were often the instigators rather than merely gatekeepers of critical viewpoints" (Althaus, 2003, p. 402). Similarly, a study by Entman (2004) revealed far more critical views of government policy in the news than predicted by the indexing hypothesis, views that were attributed to foreign sources: "Foreign sources contributed far more to the policy critique than members of Congress" (Entman, 2004, p. 55). The image of a static dependence on domestic official sources and cues on behalf of the news media has also been criticized by Lawrence, whose analyses of media reporting of police brutality showed how the dynamics of institutionally driven and event-driven news could significantly alter the opportunities for unofficial actors outside established institutions to voice their concern in the news media (Lawrence, 2000; see also Shehata, 2007; Wolfsfeld and Sheafer, 2006).

This study will investigate to what extent the indexing hypothesis explains news coverage of climate change. More specifically, we will study the indexing hypothesis by investigating the relative extent to which news coverage is focused on domestic official elite actors within established political institutions, as well as whether some issue-specific frames that have had stronger political proponents in the United States are also more prominent in American news coverage. The indexing hypothesis can, however, be pitted against other accounts of factors that shape media coverage of international issues.

Causes for Cross-national Similarity: Why Climate Change Coverage Might Not Differ

While the indexing hypothesis suggests that domestic factors in general, and domestic political elite actors and institutions in particular, have a strong influence on political news coverage, research also points to alternative factors that might work to homogenize news coverage of international issues across countries. In the case of global warming there are, in fact, several reasons why the similarities in news coverage in different countries are more striking than the dissimilarities.

First, research has shown that international news is heavily influenced by a few dominant news wires, such as AP, AFP and Reuters, that provide national news organizations with a constant flow of news from around the world (Campbell, 2004; Graber, 2006). Together with global television networks, such as CNN, BBC World, al-Jazeera, Sky News, etc., these news organizations have an important inter-media agenda-setting function for national media outlets in times when the number of foreign correspondents has shrunk for financial reasons (McCombs, 2004; Rai and Cottle, 2007; Reese and Danielian, 1987).

Second, some researchers argue that the journalistic profession is becoming ever more similar across western democracies in terms of values, norms and routines. That is, apart from the fact that the dominant sources of international news—news agencies and other leading news organizations—that domestic reporters use are limited, news organizations across the western world share a common conception of what constitutes news and how to report it, according to this argument (Hallin and Mancini, 2004). The claim of media homogeneity is central to the "new institutionalism" approach on the news media that has been advocated by several scholars (Cook, 2005; Meyer, 2002; Sparrow, 1999). However, the degree to which such homogeneity in journalistic practice and content characterizes news coverage across different countries and media types is still an open empirical question.

Third, the issue of climate change is different from many other international topics reported frequently in the news media. Its uniqueness has very much to do with the scientific consensus surrounding the issue, as well as the accessible information on the current state of knowledge about global warming—made publicly available to all news organizations through the assessment reports issued by the Intergovernmental Panel on Climate Change (IPCC). In that sense, at each point in time reporters around the world have had access to the same information when covering climate change, which makes it more likely that an issue is reported similarly by different news organizations.

Climate Change: Frames and Counter-frames

The basic research question addressed in this study is whether domestic political factors influence how climate change is reported by national news media organizations. More specifically, we are interested in the way climate change is *framed* in news coverage, as well as to what extent such framing reflects the lines of argument among the domestic political elite. As argued by several scholars, framing refers to an inevitable process of sense making in order to give structure and meaning to the world outside (D'Angelo and Kuypers, 2010; Schaffner and Sellers, 2010). Frames, on the other hand, refer to those specific organizing principles that give structure and meaning to the world outside (Gitlin, 2003; Reese, 2003). As stated by Reese: "[f]rames are *organizing principles* that are socially *shared* and *persistent* over time, that work *symbolically* to meaningfully *structure* the social world" (2003, p. 11, emphasis in original). Our conception of frames follows work done by Hall et al. (1978) and Entman (1993, 2004), who attribute specific importance to the problem-defining character of frames. According to Entman, to frame is "is to *select some aspects of a perceived reality and make them more salient in a communicating text, in such a way as to promote a particular problem definition, causal interpretation, moral evaluation, and/or treatment recommendation* for the item described" (1993, p. 52, emphasis in original). Even though several studies have analyzed news coverage of climate change, not all have done so explicitly using a framing approach. It is, however, possible to identify key frames that are relevant to our study of climate change coverage based on this research.

Studies of media representations of climate change have—in one way or another—investigated the relative salience of climate change frames and various counter (or oppositional) frames in news coverage. For instance, Boykoff and Boykoff (2004) found that news reporting in American prestige newspapers between 1988 and 2002 deviated in significant ways from the scientific consensus regarding the causes of global warming. Fifty-three percent of all stories analyzed gave a "balanced" view of anthropogenic contributions to global warming (Boykoff and Boykoff, 2004, p. 129). However, roughly 41 percent of the stories either treated human activity as the exclusive (6 percent) or dominant (35 percent) cause of global warming, as opposed to accounts treating global warming as the result of natural variations. A follow-up study of American television news revealed a similar pattern where "balanced" accounts dominated (Boykoff, 2008). It is worth noting that while the high level of "balanced" stories evident from these studies might deviate from the dominant view promoted by the IPCC, news coverage has, overall, still been supportive of the picture of human-induced global warming given by the IPCC to a substantially larger extent than the view of natural variations promoted by climate skeptics. This conclusion is supported by a comparative study of American and German news coverage of the Kyoto negotiations: "In one respect . . . the American and German media were alike: all news outlets that we examined framed global warming far more often as a real threat to humankind than as something unproven, undetermined, or beneficial" (Nacos et al., 2000).

Based on Reese's (2003) and Entman's (1993) definitions of framing, Trumbo's (1996) application to the climate change debate, as well as the studies cited above, we consider the climate change frame (defined below) to be the central organizing idea that structures the debate on global warming. This climate change frame has been challenged by several counter-frames, which are promoted by various political actors opposed to global warming legislation. Let us briefly discuss these frames before specifying the hypotheses that will guide our comparison between American and Swedish news coverage of climate change.

The climate change frame posits that global warming is a significant social problem caused by human activity through the emission of carbon dioxide and other so-called greenhouse gases into the atmosphere. Given this problem definition and causal interpretation, reduction of greenhouse gas emissions follows as the logical solution according the climate change frame. This climate change frame is embraced fully and promoted by the IPCC, at least since its second assessment report released in 1995 (Giddens, 2009; Krosnick et al., 2000). In response to the climate change frame, actors opposed to binding legislations for emission reduction have promoted several counter-frames. As McCright and Dunlap (2000, p. 506) argue, these counter-frames have been promoted with the purpose of constructing the "*non*-problematicity" of global warming. In that sense, they have developed as a direct response to the climate change frame. McCright and Dunlap's study of the American conservative movements counter-claims in the debate over global warming revealed three broad counter-frames consequently used by conservative think-tanks challenging emission reductions (McCright and Dunlap, 2000, p. 510; see also Antilla, 2005). First, a scientific-uncertainty frame claims that the scientific community is divided regarding global warming, that existing research is inconclusive when it comes to the causes and consequences of climate change, and that natural variation is the driving force behind increasing global temperatures. Consequently, more research is needed before any actions are taken (Grundmann, 2006). The scientific-uncertainty frame has been the main focus of several studies of news coverage of climate change (Antilla, 2005; Boykoff and Boykoff, 2004; Nacos et al., 2001; Zehr, 2009). Second, the economic consequences frame has been used primarily as an argument against binding legislation. Caps on emissions would substantially harm the national economy and lead to higher levels of unemployment. More recently, however, environmentalists and politicians who favor actions to combat global warming have also used an economic opportunity frame that recasts climate change as an opportunity to boost the economy (Nisbet, 2010, p. 56). Third, McCright and Dunlap (2000) also found a benefit frame that denies the negative consequences of global warming. Instead, increasing temperatures could improve quality of life, health and agriculture. Accordingly, measures to combat global warming are not necessary.

In order to assess the validity of competing claims regarding agenda- and frame-building processes, this study will analyze how climate change has been reported in American and Swedish newspapers. While the counter-frames discussed above were originally identified in an American context, some studies have confirmed their presence in European countries as well—even though they seem to be less prominent in the European media (Boykoff and Rajan, 2007; Dirikx and Gelders, 2009; Nacos et al., 2001). The next section outlines the rationale behind the research design and specifies the hypotheses that will guide our empirical analysis.

Research Design and Hypotheses

The potential "indexing" of global warming to domestic mainstream government debate will be analyzed by comparing news coverage of climate change in American and Swedish newspapers. By juxtaposing news coverage in these two countries this study intends to establish whether differences in national political debate are reflected in news coverage—a core idea of the indexing literature.

Observers have pointed to the political divide that has existed between the United States and European countries when it comes to responding to climate change. Krämer (2004), for instance, argues that political-cultural factors have accounted for differences between American and European environment policies since the mid-1980s. Central in this regard are deep-rooted ideas regarding the role of the state and the market in society, which become manifest in global environmental negotiations where the United States tends to subordinate environmental questions to economic and trade issues: "This leads it to defend interests that sometimes appear to be those of US industry, not those of the global environment" (Krämer, 2004, p. 67). Europe, on the other hand, has had a stronger commitment to social and environmental concerns—differences that became clearly evident in the negotiation over the Kyoto protocol (Krämer, 2004, p. 65). A similar political culture argument has been put forward by Schreurs:

> The differences between the EU and the US in the climate change negotiations reflect the different political cultures that have taken root in Europe and the US. Environmental protection has become a major issue for EU foreign policy, and the EU has increasingly come to see itself as an international environmental leader. (2004, p. 222)

While the Clinton administration embraced the IPCC view on global warming, the 1994 Republican victory in the congressional elections increased the divide between the White House and Congress on how to address climate change. The US Congress was skeptical of climate change science, opposed to an agreement that could affect the American economy negatively or that would exclude developing nations from any commitments (Schreurs, 2004, p. 213). That is, counter-frames, such as the scientific-uncertainty and economic consequences frame, have had strong advocates both within key political institutions as well as prominent conservative and industry groups in the United States (McCright and Dunlap, 2000; Schreurs, 2004). These frames were also adopted and promoted by President Bush at the time of the American withdrawal from the Kyoto process in 2001.

Swedish climate change policy has, on the other hand, not only been in line with the European, but also been in the frontline. Zannakis calls Swedish climate change policy "paradoxical" given the social dilemma character of global environmental politics:

> The paradox of Sweden's behaviour is that the country has probably contributed most of all countries in the climate change issue, in relation to its international commitments and GHG [greenhouse gas] emissions per capita. Sweden is unique in that the country will over-implement its international commitments, while already having among the lowest per capita GHG emissions in the industrialized world. (2009, p. 18)

Sweden was one of the promoters of a binding protocol, and even opted for more ambitious reduction targets than required by international agreements. According to Zannakis' study of Swedish climate change discourse, the official position has closely followed the climate science discourse, and—with very few exceptions—there "were never really any hesitations regarding the truthfulness of the greenhouse theory" (Zannakis, 2009, p. 104).

Based on these differences in the official political global warming debate in the two countries, the indexing hypothesis predicts a corresponding difference in climate change news coverage in the United States and Sweden. In order to test the indexing hypothesis this study addresses two research questions: (1) to what extent is climate change coverage pegged to activities taking place within domestic political institutions? And (2) to what extent is the climate change frame (as defined above) challenged in news coverage? More specifically, the indexing hypothesis leads us to expect the following:

H1: News coverage of climate change is pegged to activities taking place within domestic political institutions.

H2: News coverage of climate change is dominated by domestic official sources.

H3: The climate change frame is present in both American and Swedish news coverage.

H4: Climate change counter-frames are more prominent in American than in Swedish news coverage.

Two counter-frames will be in focus here: the scientific-uncertainty frame and the economic consequences frame. These have been crucial to domestic opponents of climate change regulation in the United States and are, therefore, relevant in light of the indexing hypothesis. Evidence that indicates stronger similarities than differences in American and

Swedish news coverage of climate change would suggest that domestication factors have less influence than international homogenization on news coverage of global warming. The next section describes the methodological strategy for testing these hypotheses.

Methodology and Data

To answer our research questions and test the hypotheses outlined above, we conducted a content analysis of climate change coverage in two prominent dailies in the United States and Sweden: *New York Times* and *Washington Post*, as well as *Dagens Nyheter* and *Svenska Dagbladet*. These outlets were selected because they are commonly considered as elite quality papers that share similar functions within each country (Dimitrova and Strömbäck, 2008, p. 209). The content analysis was conducted in two phases. First, an extensive content analysis tracked the flow of news stories published in the four papers between the Kyoto (December 1, 1997) and Bali (December 31, 2007) climate change summits. The American articles were collected from Lexis/Nexis using the search string "Kyoto" OR "climate change" OR "global warming" appearing in the leading paragraphs. Editorials, opinion pieces and stories only mentioning climate change in passing were excluded from the analysis, so that the final sample consisted of news items about climate change or global warming in particular. A similar procedure was followed in gathering the Swedish articles, using the following database search strings on Presstext and Mediearkivet: "kyoto* OR klimatför* OR växthuseffekt* OR växthusgas*". This generated 934 articles from the American papers (473 from the *New York Times* and 461 from the *Washington Post*) and 847 articles from the Swedish papers (489 stories from *Dagens Nyheter* and 358 from *Svenska Dagbladet*).

In order to investigate the general focus of American and Swedish climate change coverage during this 10-year period, each article was coded according to the triggering event to which the story was pegged (see Bennett and Livingston, 2003; Sigal, 1973; Wolfsfeld and Sheafer, 2006 for similar approaches). The purpose of this variable is to capture the extent to which news coverage is driven by—or tied to—activities or events taking place in domestic political institutions, at international summits, within the science community or among other non-governmental organizations (NGOs). Rather than being a direct measure of exactly who initiated an article, the coder was asked to judge whether the reported activity or event was triggered by actions controlled by (1) domestic official actors or institutions (the government, other politicians, public administration personnel), (2) foreign official actors or institutions, (3) climate scientists and research reports, (4) NGOs, (5) the news media themselves (through investigative journalism or thematic reporting including a variety of sources), or (6) by spontaneous events not controlled by any actor in particular, such as an accident. An intra-coder reliability check of 102 articles yielded agreement scores between 98 and 100 percent, with acceptable kappa values.[1]

Second, we also conducted an intensive comparative case study of news coverage of the Kyoto and Bali summits. Building on an analytic strategy used in past studies of indexing (Mermin, 1999; Zaller and Chiu, 1996), each summit story was coded on a paragraph-by-paragraph basis in order to capture the relative prominence of various sources and frames in the news. The *New York Times* and *Washington Post* articles were retrieved from Lexis/Nexis using the search strings "Kyoto", followed by a manual selection of all relevant news stories published between December 1 and 14, 1997, and "Bali", followed by a manual selection of all relevant stories published between December 3 and 16, 2007. Swedish articles from *Dagens Nyheter* and *Svenska Dagbladet* were collected manually from the same period. To capture the dominant theme of the stories this content analysis was based on the initial 50 percent of the paragraphs of each article. For instance, the first 12 paragraphs were coded for a 24-paragraph story, the first eight were coded for a 16-paragraph article, etc. Sources and frames were coded on a presence–absence basis for each paragraph of the news stories. A variety of source categories were included in the code sheet, such as domestic government actors, domestic opposition party actors, foreign government actors, climate scientists, NGOs, industry groups, etc. A source was deemed present whenever an actor appeared with statements, facts or quotes attributions in the paragraph.

An important property of frames is their latent character. That is, despite their presence they are seldom fully manifested in communicating texts (Entman, 1993; Reese et al., 2003). That is why scholars often analyze fragments, or indicators, of various frames in search for their presence in media coverage. This becomes obvious when news coverage of issues is characterized by strong consensus over problem definitions—which is a key feature of a frame. In such cases news coverage may be built around a common problem definition without mentioning it explicitly, but instead be focused on different causal interpretations or treatments. As this might very well be the case with respect to the climate change frame—especially in the context of institutional summits organized with the specific purpose of fighting global warming—our analysis is based on what we consider to be key fragments of the climate change frame as well as fragments of the counter-frames described above. Therefore, the climate change frame was coded as present whenever a paragraph mentioned (1) human activity as a cause of global warming (95 percent agreement, Cohen's kappa = 0.90)[2] or (2) whenever reducing the emission of heat-trapping gases was mentioned as a way of combating climate change (95 percent agreement, Cohen's kappa = 0.90). The scientific-uncertainty frame was coded as present whenever (1) human activity was explicitly questioned as a cause of global warming (100 percent agreement, kappa not calculated), (2) natural variations were mentioned as a cause of global warming (100 percent agreement, kappa not calculated) or (3) climate science/knowledge was depicted as inconclusive regarding the extent or causes of global warming (100 percent agreement, kappa not calculated). The economic consequences frame was coded as present whenever the economic consequences of binding legislation were discussed, either in general economic terms or by mentioning costs in terms of numbers (96 percent agreement, Cohen's kappa = 0.65).

Results

To give an impression of how climate change coverage has developed in the American and Swedish press during the 10-year period between January 1998 and December 2007, Figure 1 displays the amount of coverage on a monthly basis. The graphs are based on the number of paragraphs published each month rather than the number of articles in order to adjust for the substantially longer articles published in the *New York Times* and *Washington Post* compared to *Dagens Nyheter* and *Svenska Dagbladet*. A number of points are interesting to note. First, climate change seems to have been a more prominent issue in the American newspapers than in the Swedish ones during most

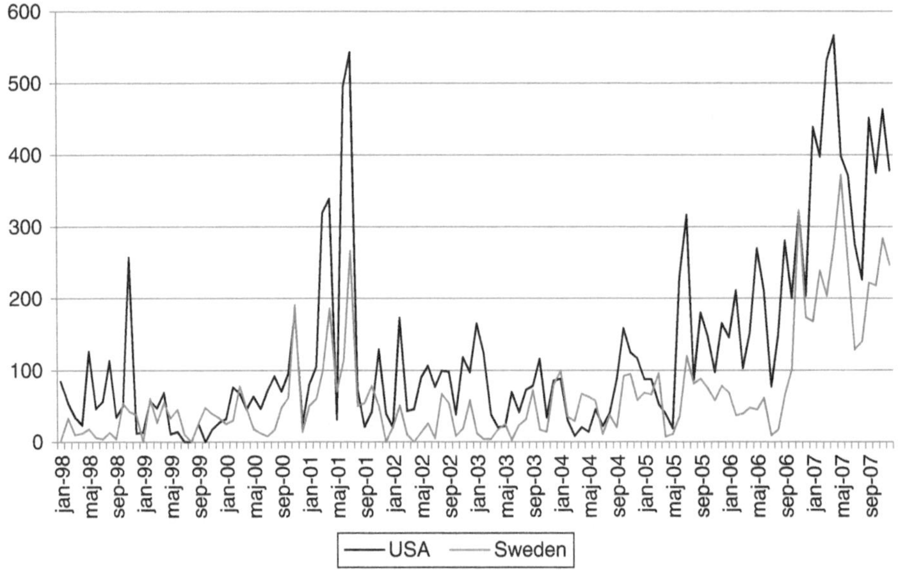

FIGURE 1

Amount of news coverage of climate change in US and Swedish newspapers (number of paragraphs)

years. This is also supported by the larger total number of American news stories (934 articles compared to 847). Second, there is a clear trend of increasing coverage over time. While there are some apparent ups and downs in the first seven-year period, 2005 and 2006 display an obvious surge in the amount of news coverage devoted to climate change in both the United States and Sweden. Third, the graph also shows significant similarities between the two countries in terms of attention dynamics—peaks and dips seem to take place in tandem. In fact, the correlation between two trends is 0.82 (Pearson's r), suggesting a rather strong relationship between the media agendas in the two countries.

Based on indexing, the first hypothesis suggested that news coverage in both American and Swedish newspapers should be pegged to activities taking place within domestic political institutions, as opposed to international organizations, foreign governments, NGOs and other unofficial actors. Table 1 shows that while domestic political actors and institutions serve as triggers to far less than a majority of the articles published during these 10 years, important differences exist between the United States and Sweden in this regard. Overall, domestic officials and institutions account for approximately 37 percent of the stories in the United States, while *Dagens Nyheter* and *Svenska Dagbladet* peg 20 and 13 percent of their stories to activities controlled by these actors. More specifically, about 30 percent of all American news stories are pegged to activities originating from the White House or controlled by members of Congress. By comparison, only 8.5 percent of the Swedish articles were pegged to activities controlled by the Swedish government or by members of Riksdagen, the Swedish parliament. Swedish climate change coverage is more often pegged to activities taking place abroad than US coverage—such as actions taken by foreign countries, the European Union and, to a lesser extent, international climate change summits. These activities account for approximately 40 percent of the Swedish stories, but only for 20–25 percent of the American stories.

TABLE 1
Main actor/institution to which news stories are pegged (%)

	New York Times	*Washington Post*	*Dagens Nyheter*	*Svenska Dagbaldet*
Domestic officials	38.2	36.0	20.2	13.2
Domestic government	18.8	15.4	5.1	3.9
Members of congress/ parliament	13.3	13.0	5.3	2.0
Public administration	6.1	7.6	9.8	7.3
Foreign officials	25.2	20.5	39.5	42.2
Foreign government/ opposition	6.6	5.0	9.2	13.1
Climate summits	12.1	12.1	15.3	16.5
United Nations/IPCC	4.6	3.0	5.0	5.9
European Union	1.9	0.4	10.0	6.7
Unofficial actors	20.6	18.9	20.4	23.7
NGO	5.2	4.6	9.8	10.6
Scientific community	15.4	14.3	10.6	13.1
Other	16.0	24.6	19.1	12.2
N (articles)	473	461	489	358

In sum, the indexing hypothesis is supported more clearly with respect to American than to Swedish news coverage. At the same time, even here domestic political institutions only account for about 30 percent of climate change coverage. The scientific community accounts for almost as many news stories as the White House does—mainly through the release of scientific reports—and the balance between news stories pegged to the scientific community, on the one hand, and domestic political institutions, on the other, is even tilted in favor of the former in the Swedish case. Thus, even though domestic political institutions seem to be one of the key triggers of climate change coverage in the United States, they do not trigger the majority of the stories.

Turning to news coverage of the Kyoto and Bali summits specifically, Table 2 presents results relevant to our second hypothesis concerning source prominence. Again, domestic official dominance seems to be a better characterization of climate change coverage in the American than in the Swedish newspapers. Twenty-two percent of all paragraphs in *New York Times* and *Washington Post* about the Kyoto summit included statements or quotes from officials working for the Clinton administration. By comparison, only about 6 percent of the Swedish paragraphs covering this period included statements or quotes from Swedish government sources. Swedish coverage relies mainly on sources from abroad—foreign governments (including the United States), the European Union as well as United Nations and IPCC actors. This reliance on foreign actors, organizations and institutions prevails in the Swedish press coverage of the Bali conference, even though domestic government sources have gained some prominence. Domestic government and congress sources have lost prominence in the American newspapers though, while foreign actors representing governments of other countries, United Nations and the IPCC are present more often. In terms of relative prominence, White House representatives are still the most frequent source category in the *New York Times* and *Washington Post*.

There are, however, some important differences between newspapers within the two countries. In the Kyoto case, for instance, White House sources are more prominent in

TABLE 2
The use of sources in summit coverage (%)

	Kyoto 1997		Bali 2007	
	United States	Sweden	United States	Sweden
Domestic official actors	31.4	5.9	14.4	9.6
Domestic government	22.0	5.9	12.0	9.3
Members of congress/parliament	9.4	0	2.4	0.6
Public administration source	0	0	0	0
Foreign official actors	5.3	17.7	15.0	17.9
Foreign government	2.3	7.4	5.3	11.1
United Nations/IPCC	1.8	4.4	8.7	5.6
European Union	1.2	5.9	1	1.2
Unofficial actors	9.4	0	4.8	5.2
NGO	3.5	0	3.8	4.3
Business/industry	4.7	0	1.0	0.9
Scientists	1.2	0	0	0
N (paragraphs)	341	135	208	324

the *New York Times* (28.7 percent) than in the *Washington Post* (17.6 percent), while the opposite is true when it comes to Congress sources, who appear in 4.4 percent of the paragraphs in the *New York Times* and in 12.7 percent in the *Washington Post.* Furthermore, during the Bali conference foreign government sources all appear in the *Washington Post* (10 percent) while being almost totally absent in the *New York Times* (0.9 percent). Similarly, all NGO and interest group sources in the American coverage of the Kyoto summit appear in the *Washington Post* (5.9 percent).

Overall, the data lends partial support for the second hypothesis. Domestic political elite actors dominate as sources to a larger extent in the United States than in Sweden, where foreign government sources appear more frequently. At the same time, White House representatives and Congress members are accompanied by a broader range of unofficial sources such as NGOs, business and industry groups as well as scientists in the case of Kyoto coverage.

Finally, Table 3 presents data that speak to hypothesis 3 and 4, regarding the presence of the climate change frame as well as the two counter-frames. Hypothesis 3 predicted the climate change frame to be present in both American and Swedish news coverage. This is certainly the case, as can be seen from Table 3. The climate change frame is present in between 50 and 79 percent of the paragraphs. To reiterate, the presence of

TABLE 3
The presence of the climate change frame, the scientific-uncertainty frame and the economic consequences frame (%)

	Kyoto 1997		Bali 2007	
	United States	Sweden	United States	Sweden
Climate change frame	67.4	50.4	72.1	79.0
Scientific-uncertainty frame	0.6	0	0	0.3
Economic-consequences frame	21.1	7.4	0.5	4.3
N (paragraphs)	341	135	208	324

the climate change frame is based on two indicators: whether the paragraph mentions (1) human activity as a cause of global warming or (2) whether reducing the emission of heat-trapping gases was mentioned as a way of combating climate change. These numbers point to a widespread use of the climate change frame in American and Swedish newspapers.[3] However, the presence of this frame is only interesting in relation to the two counter-frames—only in relation to these competing frames are we able to judge the pervasiveness of the climate change frame in the media.

Hypothesis 4 predicted the two counter-frames to be more prominent in the American than in the Swedish newspapers. This hypothesis is only partially supported by the data. First, the total absence of the scientific-uncertainty frame is both striking and a bit surprising given past research on climate change coverage in the United States. We will return to this finding in the discussion. Second, the economic consequences frame is particularly prominent in American news coverage of the Kyoto summit—where it appears in 21.1 percent of the paragraphs. The economic consequences frame has disappeared almost completely in 2007 though, present only in 0.5 percent of the paragraphs. Its presence in the Swedish newspapers shrinks from 7.4 percent of the paragraphs at the Kyoto summit to 4.3 at the Bali summit. Overall, however, the results indicate that the climate change frame clearly dominates press coverage, and this dominance increases over time.

Discussion

Research on news coverage of climate change has repeatedly pointed to the ways the media "distorts" the consensus view on climate change within the scientific community in general, and the IPCC in particular. For example, Boykoff and Boykoff's (2004) widely cited study of American climate change coverage between 1998 and 2002 concludes that "the US prestige press—the *New York Times*, the *Washington Post*, the *Los Angeles Times*, and the *Wall Street Journal*—has contributed in significant ways to this failed discursive translation through the adherence to journalistic norms, and more specifically to the journalistic norm of balance. In the end, adherence to the norm of balanced reporting leads to informationally biased coverage of global warming" (Boykoff and Boykoff, 2004, p. 134). This general view of climate change reporting is echoed in several subsequent scholarly accounts of media coverage and global warming, even though there is a sense of growing media reflection of the scientific consensus in later years (Antilla, 2005; Boykoff, 2005; Hulme, 2009; Lewis and Boyce, 2009; Zehr, 2009). Crucial in this regard has been the scientific-uncertainty frame promoted by various climate skeptics as a way of fending off the climate change as a problem (Antilla, 2005; Boykoff and Boykoff, 2004; McCright and Dunlap, 2000; Trumbo, 1996; Zehr, 2000). The findings of this study deviate from some of these claims about climate change coverage, and offer a slightly different view of how the news media have reported on global warming. Let us, therefore, briefly discuss our main findings as well as their implications for research on news coverage of climate change and press–politics relations.

This paper has approached the question of climate change coverage from a comparative perspective, using the indexing hypothesis as the theoretical point of departure. In that sense, this is one of the first studies that actually analyzes climate change coverage comparatively in a systematic manner (see Nacos et al., 2001 for an exception). The premise of this country comparison was based on the indexing hypothesis. More specifically, we expected that differences in the national political environment surrounding the climate change issue in the United States and Sweden would be reflected in news coverage. On the other hand, it was argued, international news flows and inter-media agenda-setting processes, coupled with the unique presence of an international scientific body founded on the purpose of delivering climate reports in a consensual manner, are factors likely to streamline coverage of global warming across countries. Climate change coverage in the United States and Sweden was analyzed against these two competing perspectives on global news. In sum, the results showed that news coverage of climate change has developed quite similarly in both countries between 1998 and 2007. Together with the finding that an overwhelming majority of these news stories were pegged to activities not controlled by domestic political institutions, this indicates that international factors have played a more prominent role in climate change coverage than is suggested by the indexing hypothesis. However, the American news stories were more often pegged to activities taking place within major domestic political institutions, and the *New York Times* and *Washington Post* also relied more on domestic politicians as sources in coverage of the Kyoto and Bali summits than the Swedish newspapers.

Furthermore, the intensive case study of coverage during the Kyoto and Bali summits revealed a striking dominance of the climate change frame in both the American and Swedish newspapers. The view that global warming is a significant social problem caused by human activity permeates news reporting in both countries at both summits—a finding that supports our third hypothesis. However, the fourth hypothesis was only partially supported. Despite attempts by various political actors in the United States to question climate change policy, their crucial frame—the scientific-uncertainty frame—was totally absent in news coverage of the two summits. On the other hand, the counter-frame hypothesis was supported by the clear presence of the economic consequences frame in the US Kyoto coverage. Only in this regard did the framing of climate change differ between the American and the Swedish papers in a way that reflects the different political environments in the two countries.

How can we explain the total absence of the scientific-uncertainty frame in the news media? Both the presence of influential frame promoters within the American political elite as well as past research on US coverage of climate change strongly suggest that voices skeptical to anthropogenic climate change are prominent in the media—especially in the *New York Times* and *Washington Post* coverage in the years leading up to and following the Kyoto summit (Boykoff and Boykoff, 2004; McCright and Dunlap, 2000). Just to reiterate, according to the findings of this study, the climate change frame strongly dominates American news coverage of global warming and leaves no room for climate skeptics questioning the validity of the IPCC view—a conclusion that appears to be totally at odds with past research.

These results are most likely explained by factors related to the design of this study. The pivotal study by Boykoff and Boykoff (2004) analyzed routine coverage in the American prestige press between 1988 and 2002, while the follow-up study was based on television news between 1995 and 2004 (Boykoff, 2008). Routine times might be substantially different than summit coverage. Climate summits entail a very specific news-reporting context where the problem definitions are clearly institutionally defined.

These summits are deliberately set up with a common purpose: to arrive at a common strategy to fight global warming. On the other hand, while this consensual political environment might explain the omnipresence of the climate change frame, it does not determine the exclusion of counter-frames supported domestically. Perhaps the findings of this study point to how contextual factors influence the framing of climate change in the news media. In situations where frames are clearly institutionally defined, counter-frames are less likely to gain prominence in the news, despite the presence of oppositional voices within domestic official circles.

NOTES

1. For the variables used here, the reliability scores were: domestic government, 100 percent agreement; other politicians, 99 percent agreement, Cohen's kappa = 0.95; public administration, 98 percent agreement, Cohen's kappa = 0.91; foreign government, 100 percent agreement; summits, 99 percent agreement, Cohen's kappa = 0.94; United Nations, 99 percent agreement, Cohen's kappa = 0.66; European Union, 99 percent agreement, Cohen's kappa = 0.93; NGOs, 99 percent agreement, Cohen's kappa = 0.80; scientists, 98 percent agreement, Cohen's kappa = 0.89.
2. Based on an intra-coder test of 80 randomly selected paragraphs.
3. Differences between newspapers within both countries exist with regard to the climate change frame. With respect to the Kyoto summit: *New York Times* 55.9 percent, *Washington Post* 75.1 percent, *Dagens Nyheter* 39.7 percent and *Svenska Dagbladet* 64.9 percent. With respect to the Bali summit: *New York Times* 73.1 percent, *Washington Post* 71.0 percent, *Dagens Nyheter* 83.0 percent and *Svenska Dagbladet* 74.8 percent. However, we do not consider these differences to be important in their own right as the climate change frame still permeates news coverage at both summits. The relevant question is rather whether—and to what extent—this frame is challenged by counter-frames.

REFERENCES

ALTHAUS, SCOTT (2003) "When News Norms Collide, Follow the Lead: new evidence for press independence", *Political Communication* 20, pp. 381–414.

ANTILLA, LIISA (2005) "Climate of Skepticism: US newspaper coverage of the science of climate change", *Global Environmental Change* 15, pp. 338–52.

BENNETT, LANCE W. (1990) "Toward a Theory of Press–State Relations", *Journal of Communication* 40, pp. 103–25.

BENNETT, LANCE W., LAWRENCE, REGINA G. and LIVINGSTON, STEVEN (2007) *When the Press Fails. Political power and the news media from Iraq to Katrina*, Chicago: University of Chicago Press.

BENNETT, LANCE W. and LIVINGSTON, STEVEN (2003) "Gatekeeping, Indexing, and Live-event News: is technology altering the construction of news?", *Political Communication* 20, pp. 359–62.

BENNETT, LANCE W. and PALETZ, DAVID L. (1994) *Taken by Storm: the media, public opinion, and U.S. foreign policy in the Gulf War*, Chicago: University of Chicago Press.

BOYCE, TAMMY and LEWIS, JUSTIN (2009) *Climate Change and the Media*, New York: Peter Lang.

BOYKOFF, MAXWELL T. (2008) "Lost in Translation? United States television news coverage of anthropogenic climate change, 1995–2004", *Climate Change* 86, pp. 1–11.

BOYKOFF, MAXWELL T. and BOYKOFF, JULES M. (2004) "Balance of Bias: global warming and the US prestige press", *Global Environmental Change* 14, pp. 125–36.

BOYKOFF, MAXWELL T. and RAJAN, RAVI S. (2007) "Signals and Noise: mass media coverage of climate change in USA and the UK", *European Molecular Biology Organization Reports* 8, pp. 207–11.

CAMPBELL, VINCENT (2004) *Information Age Journalism. Journalism in an international context*, London: Arnold.

COOK, TIMOTHY (2005) *Governing With the News. The news media as a political institution*, Chicago: University of Chicago Press.

D'ANGELO, PAUL and KUYPERS, JIM A. (2010) *Doing News Framing Analysis. Empirical and theoretical perspectives*, New York: Routledge.

DIMITROVA, DANIELA and STRÖMBÄCK, JESPER (2008) "Foreign Policy and the Framing of the 2003 Iraq War in Elite Swedish and US Newspapers", *Media, War and Conflict* 1(2), pp. 203–20.

DIRIKX, ASTRID and GELDERS, DAVE (2009) "Global Warming Through the Same Lens: an explorative study framing study in Dutch and French Newspapers", in: Tammy Boyce and Justin Lewis (Eds), *Climate Change and the Media*, New York: Peter Lang.

ENTMAN, ROBERT (1993) "Framing: towards clarification of a Fractured Paradigm", *Journal of Communication* 43, pp. 51–8.

ENTMAN, ROBERT M. (2004) *Projections of Power. Framing news, public opinion, and U.S. foreign policy*, Chicago: University of Chicago Press.

GANDY, OSCAR H. (1982) *Beyond Agenda Setting: information subsidies and public policy*, Norwood, NJ: Ablex Publishing Company.

GIDDENS, ANTHONY (2009) *The Politics and Climate Change*, Oxford: Polity.

GITLIN, TODD (2003) *The Whole World Is Watching. Mass media in the making and unmaking of the New Left*, Berkeley: University of California Press.

GRABER, DORIS A. (2006) *Mass Media and American Politics*, 7th edn, Washington, DC: CQ Press.

GRUNDMANN, REINER (2006) "Ozone and Climate: scientific consensus and leadership", *Science, Technology and Human Values* 31, pp. 73–101.

HALL, STUART, CRICHTER, CHARLES, JEFFERSON, TONY, CLARKE, JOHN and ROBERTS, BRIAN (1978) *Policing the Crisis. Mugging the state, law, and order*, London: Macmillan.

HALLIN, DANIEL C. and MANCINI, PAOLO (2004) *Comparing Media Systems. Three models of media and politics*, New York: Cambridge University Press.

HULME, MIKE (2009) "Mediated Messages about Climate Change: Reporting the IPCC Fourth Assessment in the UK Print Media", in: Tammy Boyce and Justin Lewis (Eds), *Climate Change and the Media*, New York: Peter Lang.

KRÄMER, LUDWIG (2004) "The Roots of Divergence: a European perspective", in: Norman J. Vig and Michael G. Faure (Eds), *Green Giants*, Cambridge, MA: MIT Press.

KROSNICK, JON A., HOLBROOK, ALLYSON L. and VISSER, PENNY S. (2000) "The Impact of the Fall 1997 Debate About Global Warming on American Public Opinion", *Public Understanding of Science* 9, pp. 239–60.

LAWRENCE, REGINA (2000) *The Politics of Force. Media and the construction of police brutality*, Berkeley: University of California Press.

LEWIS, JUSTIN and BOYCE, TAMMY (2009) "Climate Change and the Media: The Scale of the Challenge", in: Tammy Boyce and Justin Lewis (Eds), *Climate Change and the Media*, New York: Peter Lang.

MCCOMBS, MAXWELL (2004) *Setting the Agenda: the mass media and public opinion*, Oxford: Polity.

MCCRIGHT, AARON M. and DUNLAP, RILEY E. (2000) "Challenging Global Warming as a Social Problem: an analysis of the conservative movement's counter-claims", *Social Problems* 47, pp. 499–522.

MERMIN, JONATHAN (1999) *Debate War and Peace. Media coverage of U.S. intervention in the post-Vietnam era*, Princeton, NJ: Princeton University Press.

MEYER, THOMAS (2002) *Media Democracy. How the media colonize politics*, Cambridge: Polity Press.

MOLOTCH, HARVEY and LESTER, MARILYN (1974) "News as a Purposive Behaviour: on the strategic use of routine events, accidents, and scandals", *American Sociological Review* 39, pp. 101–12.

NACOS, BRIGITTE L., SHAPIRO, ROBERT Y., HRITZUK, NATASHA and CHADWICK, BRUCE (2001) "New Issues and the Media: American and German news coverage of the global-warming debate", in: Brigitte Nacos, Robert Y. Shapiro and Pierangelo Iserina (Eds), *Decisionmaking in a Glass House*, Lanham, MD: Rowman and Littlefield.

NISBET, MATTHEW C. (2010) "Knowledge into Action: framing the debates over climate change and poverty", in: Paul D'Angelo and Jim A. Kuypers (Eds), *Doing News Framing Analysis*, New York: Routledge.

RAI, MUGDHA and COTTLE, SIMON (2007) "Global Mediations on the Changing Ecology of Satellite Television News", *Global Media and Communication* 3, pp. 51–78.

REESE, STEPHEN D. (2003) "Prologue—framing public life: a bridging model for media research", in: Stephen D. Reese, Oscar H. Gandy and August E. Grant (Eds), *Framing Public Life. Perspectives on media and our understanding of the social world*, Mahwah, NJ: Lawrence Erlbaum Associates.

REESE, STEPHEN D. and DANIELIAN, LUCIG H. (1987) "Intermedia Influence and the Drug Issue: converging on cocaine", in: Pamela Shoemaker (Ed.), *Communication Campaigns About Drugs. Government, media, and the public*, Hillsdale, NJ: Lawrence Erlbaum.

REESE, STEPHEN D., GANDY, OSCAR H. and GRANT, AUGUST E. (Eds). (2003) *Framing Public Life. Perspectives on media and our understanding of the social world*, Mahwah, NJ: Lawrence Erlbaum Associates.

RIEGERT, KRISTINA (1998) *Nationalizing Foreign Conflict*, Stockholm: Department of Political Science, University of Stockholm.

SCHAFFNER, BRIAN F. and SELLERS, PATRICK, J. (2010) *Winning With Words. The origins and impact of political framing*, New York: Routledge.

SCHEUFELE, DIETRAM A. (1999) "Framing as a Theory of Media Effects", *Journal of Communication* 49, pp. 103–22.

SCHOENBACH, KLAUS, DE RIDDER, JAN and LAUF, EDMUND (2001) "Politicians on TV News: getting attention in Dutch and German election campaigns", *European Journal of Political Research* 39, pp. 519–31.

SCHREURS, MIRANDA A. (2004) "The Climate Change Divide: the European Union, the United States, and the future of the Kyoto Protocol", in: Norman J. Vig and Michael G. Faure (Eds), *Green Giants, Cambridge*, MA: MIT Press.

SHEHATA, ADAM (2007) "Facing the Muhammad Cartoons: official dominance and event-driven news in Swedish and American elite press", *The Harvard International Journal of Press/Politics* 12(4), pp. 131–53.

SHOEMAKER, PAMELA J. and REESE, STEPHEN D. (1996) *Mediating the Message. Theories of influences on mass media content*, White Plains, NY: Longman.

SIGAL, LEON V. (1973) *Reporters and Officials. The organization and politics of newsmaking*, Lexington, MA: Heath.

SPARROW, BARTHOLOMEW H. (1999) *Uncertain Guardians. The news media as a political institution*, Baltimore, MD: Johns Hopkins University Press.

TRUMBO, CRAIG (1996) "Constructing Climate Change: claims and frames in US news coverage of an environmental issue", *Public Understanding of Science* 5, pp. 269–83.

WOLFSFELD, GADI and SHEAFER, TAMIR (2006) "Competing Actors and the Construction of Political News: the contest over waves in Israel", *Political Communication* 23, pp. 333–54.

ZALLER, JOHN and CHIU, DENNIS (1996) "Government's Little Helper: U.S. press coverage of foreign policy crises, 1945–1991", *Political Communication* 13, pp. 385–405.

ZANNAKIS, MATHIAS (2009) *Climate Policy as a Window of Opportunity. Sweden and global climate change*, Göteborg: Department of Political Science.

ZEHR, STEPHEN C. (2000) "Public Representations of Scientific Uncertainty About Global Climate Change", *Public Understanding of Science* 9, pp. 85–103.

ZEHR, STEPHEN C. (2009) "An Environmentalist/Economic Hybrid Frame in US Press Coverage of Climate Change, 2000–2008", in: Tammy Boyce and Justin Lewis (Eds), *Climate Change and the Media*, New York: Peter Lang.

MEDIA REPRESENTATIONS OF CLIMATE CHANGE IN THE ARGENTINEAN PRESS

María Teresa Mercado

The media play an essential role in the construction of social reality and consequently knowledge of what is reported in the media is essential to understanding social attitudes towards significant issues such as climate change. The assumption here is that more and/or better information fosters a clearer understanding of ecological issues, environmental awareness in society, and contributes to a transformation of values, attitudes and behavior. Few authors focus on the role of journalism in the mass communication of climate change in Latin American countries; however, these countries are very important in international negotiations because they possess most of the natural reserves to be protected at this crucial time, the "post-Kyoto" stage. For these reasons, Argentina is chosen as the focus of this study.

Introduction

All international treaties emphasize the importance of communication, access to information, education and training, and public awareness in achieving a sustainable society. In the Earth Summit held in Rio de Janeiro in 1992, an explicit appeal was made to the media to become involved in information work for the preservation of the environment and in changing public attitudes. Although from the professional perspective the main role of journalists and the media is to inform, not to educate or change attitudes, the selection and reporting of information—in short, journalistic interpretation—has some undoubted influence on the perceptions of citizens especially when it concerns a field with which citizens have little or no direct contact (Nelkin, 1990).

Areas unrelated to journalism, like environmental education, also signal the importance of communication media in this work. As Novo argued,

> there is an awareness of the enormous potential not only informative, but also educational of audio-visual or written messages; the mass dissemination of news of major environmental impact should lead media professionals to exercise extreme sensitivity and care when carrying out their work, in such a way that they are able to convey the seriousness of environmental problems to the general public without falling into scaremongering or lazy anecdotes, while at the same time presenting options and alternatives to help resolve the crisis. (2003, p. 222)

The complexity of environmental problems, both in the clarification of causes and the explanation of consequences, demands a responsible attitude from reporters: an in-depth journalism based on thorough documentation, background narrative, consequences and actors involved, analysis of environmental facts and, finally, the situation of events in their proper contexts. For the reporter Víctor Bachetta (2002),

> environmental journalism is one of the most extensive and complex genres of journalism, with five points to consider: it is investigative journalism, it is a form of science journalism, it is educative journalism, it is aware of the need to fulfil a specific social responsibility, and it must be exercised with professionalism, objectivity and responsibility but without confusing this with environmental activism.

In short, environmental protection has become a universal goal and climate change is the central issue. Today, the economic crisis since 2008 has driven down the news value of climate change at a time when "according to the consideration of scientists, the problem does not lie in the finding of irreversible climate change, but in the implementation of global policies mitigating environmental degradation" (Díaz Nosty, 2009).

Research Context

There has been a steady increase in the coverage of climate change in the media: from less than 700 references in Western Europe and North American English-language newspapers in 1988 to 7500 by 2006. In the United Kingdom, coverage quadrupled in quality newspapers in 2006 across the three years since 2003, prompted by among other things, Al Gore's successful documentary and the release of the "Stern Review" (Boykoff and Roberts, 2007, p. 6).

One of the main approaches to researching climate change reports in Western media, especially in US newspapers across the 1990s, focused on how climate sceptics achieved a special place in media reports to offer arguments counter to the prevailing view of scientists who posited the existence of anthropogenic climate change. Some journalistic norms and traditions, specifically the journalistic commitment to balanced coverage, inadvertently promote sceptics' access to the media. Reporting differing views in the interests of journalistic neutrality, served to amplify the voice of a minority denying human responsibility for climate change in the prestige press (Boykoff and Boykoff, 2004). In this case, the coverage of climate change as a conflict between two opposing positions, despite the increasing consensus among climate scientists on global warming, undermined knowledge, as in many issues related to science.

Faced with the scientific uncertainty generated by US and British media, the issue expanded beyond the scientific field in other European countries and became anchored in international relations. The two antagonists are now the European Union, leader in the fight to mitigate the effects of climate change, and the United States, which continues to hinder international negotiation (Brossard et al., 2004, p. 364; Olausson, 2009, pp. 425–6).

Another key aspect has been the focus on "catastrophe" in media messages. European research suggests that reporting tends to be alarmist, steeped in the language of fear, catastrophe and disaster (Anderson, 2009, p. 174). In the German press since the mid-1980s, for example, "the media ignored scientific complexities and uncertainties and transformed them into a sequence of events leading to catastrophe and requiring immediate action" (Weingart et al., 2000, p. 280). Since 2000, journalists have increasingly made links between dramatic weather events (such as floods and heat waves) and global climate change (Carvalho and Burgess, 2005). According to Ladle et al. (2005, p. 231), "the careful, measured language of science is not well suited to the sound-bite sensationalism that is the typical mode of communication of most of the contemporary news media".

However, showing that news media are complex, highly differentiated and occupy their own particular market niches (Anderson, 1997), Carvalho (2008, pp. 149–50), in an investigation of the discourse on climate change in Portuguese media between 2003 and 2007, concludes that scaremongering does not define the coverage of climate change in the press. The same does not happen on television, for example, mainly because of the tendency to show images of natural disasters when reporting climate change.

Boykoff and Roberts (2007), in their analysis of United Nations climate change coverage in 40 English-language newspapers in 13 countries across five continents, highlight the clear need for the analysis to be extended to other countries. The only Latin American country analyzed in their study was Honduras. Recently, Gordon et al. (2010) have analyzed newspaper coverage and the presence of frames, solutions, and conflicts surrounding the global warming issue in the Mexican newspaper *Reforma*. This is also one of the newspapers chosen by the University of Hamburg to measure media attention to climate change over a 15-year period in 23 countries worldwide as part of the project Global Media Map on Climate Change. The other Spanish-language newspaper included in the sampling is *El País* (Spain). The research shows that media coverage of climate change has slowly increased from 1996, in particular in Western countries after 2004 around the publication of the Intergovernmental Panel on Climate Change (IPCC) report in 2007 and the Copenhagen Summit in 2009 (Schäfer et al., 2011).

Methodology

This article analyzes the coverage dedicated to climate change in the quality press in Argentina and identifies the news frames employed by journalists, to distinguish between different representations of climate change. The hypothesis is that there exists a media construction of climate change in Argentina distinct from the dominant frames in media reports in industrialized countries.

The technique used is a hybrid of content analysis and qualitative textual analysis to uncover the underlying cognitive context in which climate change information is reported in the Argentinean press. In an early account, Berelson (1952, p. 18) defined quantitative content analysis as "a research technique for the objective, systematic, and quantitative description of the manifest content of communications". Over time, other authors have extended the concept by introducing qualitative analysis. Thus, content analysis allows inferences through systematic and objective identification of specific features within a text:

> Content analysis is the set of interpretive procedures of communication products (messages, texts, speeches), based on measurement techniques, sometimes quantitative (statistics based on unit counts), sometimes qualitative (logic based on the combination of categories), which allows the processing of relevant data on the very conditions that have produced those texts. (Piñuel, 2002, p. 7)

In this study, a descriptive content analysis is used to list and quantify aspects such as section, genre, sources and themes in journalistic reports. The results, together with the sustained and critical reading of texts, allows the identification of the cognitive frames of the news, "leading to a strategic selection and presentation of text analyzed as evidence for the general argument" (Fürsich, 2009, p. 240).

During the last two decades there has been a steady increase in studies using frames as a basic conceptual tool although it remains "an ambiguous theoretical conceptualization, which refuses to be subsumed together under a unified label" (Vicente and López, 2009, p. 15). Most agree, however, that framing theory implies that information content not only sets the public agenda, but also implicitly forms a way of thinking about certain issues through news frames (Igartua and Humanes, 2004). But while we assume that these frames do affect the public perception (De Vreese, 2003) and the media are powerful agents in the framing processes of social discourses (Sádaba, 2001, p. 152), the analysis here stops ahead of considering the more controversial consideration of the impact of media messages. This study focuses solely on media coverage and the analysis of frames.

The framing of an issue promotes a particular definition of a theme, selecting some aspects of a perceived reality and making them more prominent in the text, delivering serious or non-serious keys to situations or framing the story in one or another section that favors one or another social reading. The framing perspective recognizes the ability of a text to define a situation or issue and establish the terms of debate (Tankard, 2001, p. 96). To frame, according to Entman (1993, p. 52), is to "select some aspects of a perceived reality and make them more salient in a communicating text, in such a way as to promote a particular problem definition, causal interpretation, moral evaluation, and/or treatment recommendation". Gitlin (1980) identifies some typical framing strategies: the overuse of officials as sources, the trivializing actions of opposition groups, the focus on events instead of larger issues, and an emphasis on one side of the argument. For Scheufele (1999, p. 112) the factors that influence how journalists frame issues include: social norms and values, organizational pressures, pressures from interest groups, journalistic methods and the ideological orientation of journalists.

Journalists may also present news using easy-to-understand interpretive packages or frames to reduce the complexity of issues for their readers/audiences. In addition to reducing complexity, however, these frames also serve as interpretive shortcuts for audience members, leading them to make attributions of responsibility or other judgments, based on different frames or interpretations offered by mass media for the same factual content (Kim et al., 2002, p. 8).

Similarities between framing and agenda setting by McCombs and Shaw (1972) have been discussed since several researches began to explore a second-level agenda setting or attribute agenda setting that examines the influence of attribute salience, or the properties, qualities and characteristics that describe objects or people in the news and the tone of those attributes (Wu and Coleman, 2009, p. 776). Thus, the transmission of the relevance of issues is the traditional agenda effect, at a first level, while the transmission of the relevance of attributes constitutes a second level (McCombs et al., 1997).

In this project, the goal is the identification of the frames as the organizing idea for news content through selection, emphasis, exclusion and elaboration; to cite Kim et al. (2002, p. 10), the "terminological or semantic differences" in how an issue is described in the media. This interpretive package or "generic frame" was defined by Semetko and Valkenburg (2000, p. 95), who found three such generic frames in the analysis of news coverage of the media representations of European political leaders and their actions: conflict, human interest and economic consequences.

In order to undertake a content analysis, the extent of the sample must first be determined which, in this case, potentially includes all articles with references to climate change in the Argentinean press. Since this sample would be unmanageably large, a representative sample of the quality press comprising the two leading newspapers in Argentina which constitute the major actors in setting the agenda were selected for study: *Clarín* and *La Nación*.

Clarín was founded in 1954 and according to the Instituto de Verificación de Circulaciones (IVC), sold an average 528,000 copies daily between 1991 and 2001, but by June 2010 these figures had fallen to 300,837 copies on Monday to Saturday and 634,376 on Sundays. *La Nación* was founded in 1870 by Bartolome Mitre, a soldier, journalist, writer and politician, who was president of Argentina. Its circulation is much lower than that of *Clarín*: in June 2010, 152,433 copies on Monday to Saturday and 280,720 on Sundays. Historically it is considered to be politically conservative, economically liberal, and Catholic (Pizarro, 2008, pp. 77–80).

The analysis period is the year from October 1, 2009 to September 30, 2010 to include the coverage before, during and after the Copenhagen Summit in December 2009. The units of analysis, identified via a keyword search (including the terms "climate change" and "global warming") in the archives of the online editions of both newspapers, included all news texts and opinion articles. The reading of these allows us to separate opinion articles signed by prominent figures within the international arena that are published in the news pages but not the opinion section. To identify the generic frames we only analyze news, given that the conditions that determine the news frame, as we have seen, do not involve opinion articles that are, for the most part, submitted by external contributors. Also excluded from the sample are letters from readers and references in cultural listings.

A first review of the articles revealed that in many news reports the reference to climate change was an isolated or modest reference, perhaps a brief mention included in an introductory paragraph with no subsequent references. In order to ensure that the sample used to determine the news frame consisted only of relevant articles, news texts were classified and ranked according to the extent to which they focused on climate change.

Reports in which climate change was mentioned briefly as only one of the challenges facing the world today were classified as Type III reports. An example here is a report about Spain assuming the rotating presidency of the European Union, where a journalist at *Clarín* wrote about the tasks for the new President: "Among the priorities—not forgetting issues such as the economic crisis, climate change, energy security and the management of migration flows" (Martín Pérez, *Clarín*, 1 January 2010).

In other reports where climate change features more prominently but is still not the main subject, or is not given more than a paragraph or a question/answer in an interview, the reports are labelled Type II (one topic among many). This type of report usually relates to the environment in general. An example here is an extensive report on the ozone layer: "Now we see that there is a relation between the thinning of the ozone layer and climate change," adds Canziani: "Between 2000 and 2010 an evaluation was initiated of how the ozone hole is responding to the stabilization of CFCs and how it affects climate change, and vice versa" (Camps, *Clarín*, 30 June 2010). Finally, there were news reports in which climate change was the primary focus and information object, the subject of the report and the main theme; these were classified as Type I (main topic).

Entman (1993, p. 52) explained that frames in the news can be examined and identified by the presence or absence of certain keywords, stock phrases, stereotyped images, sources and sentences that provide thematically reinforcing clusters of facts or judgments. Paying particular attention to headlines, subheads and lead paragraphs, the textual analysis of Type I reports allows identification of "story focus, salient words within texts, background information, discussion of underlying problem and solutions offered" (Kothari, 2010, p. 214) which reveal the frames.

Following Igartua (2006, p. 204), a codebook was compiled that included the following variables in each unit collected: date of publication, section, headline, genre, authorship (signature), sources, thematic categories, type (I, II or III) and generic frame (only news Type I).[1]

Results

Across the sample period from October 2009 to October 2010, 196 articles from *Clarín* and 290 from *La Nación* were analyzed; in aggregate a sample of 486 news items. All were included in the quantitative content analysis; a sub-sample of 186 news articles was included in the textual analyses to identify generic frames. We begin with results from the newspaper *Clarín* since it enjoys the higher circulation in Argentina.

Coverage

The 196 reports referencing climate change in *Clarín* reveal a high point in December coinciding with the Fifteenth Conference of the Parties to the United Nations Convention on Climate Change, known as COP15 or the Copenhagen Summit (Figure 1).

Published reports alluding to the summit were published from October. Gustavo Sierra, a special correspondent, was in Denmark to cover events organized by Project Sindycate, bringing together world-renowned intellectuals, journalists and opinion leaders to promote the exchange of ideas and to pressure politicians who had not yet defined

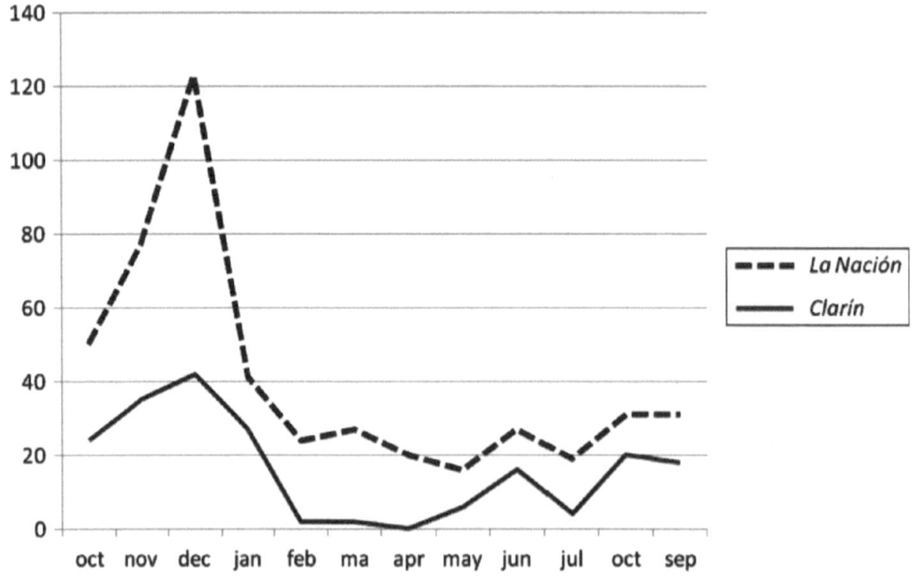

FIGURE 1
Climate change in the Argentinean press, October 2009 to September 2010

their position to support reducing emissions and sign a global climate agreement. The "Sunday Theme" of October 10 is Pollution and Development. Sierra (2009a) files a story titled: "Planet in Danger: the last chance to curb climate change". The mood of the report is dramatic. Authorities negotiate against the clock to end pollution emissions. Two months before the Copenhagen summit there is still no agreement.

The information on previous meetings of COP15 and the statements of politicians begin to monopolize space in the newspaper. On Sunday December 6, the day before the opening of the meeting, the special correspondent says in the headline of his story that the summit will "hopefully" start the next day (Sierra, 2009b). The 12-day summit was closely reported by Sierra who published daily reports except for Friday 18, the closing day. The meeting ends later than expected without agreement. On Sunday, *Clarín* takes stock of the failure.

The impact of COP15 on editorial is felt during the next month. In January we see that the number of journalistic pieces alluding to climate change remains high (27). The fall in February is pronounced: the conference is mentioned only twice. A trend that continues in March intensifies in April, the month in which no article appears in the pages of *Clarín*. In May there are six references, but not until June does the issue return to editorial salience. Gustavo Sierra is again special correspondent, this time in Berlin, where again the problems of reaching an agreement are manifest among the participating nations in international negotiations. Funding appears to be crucial. During the month of August, *Clarín* again considers climate change to be a topic of interest in its thematic agenda. The wave of natural disasters (extreme heat and fires in Russia, floods in Pakistan) and the World Meteorological Organization report reveals that 2010 is the warmest year since 1830, which led to climate change being on the front page of the Sunday 22 issue (opening section double page), accompanied by a considerable display of computer graphics. In September, the number of references to the subject remains close to the twenties and it can be assumed that as the next World Conference, COP16, approaches, the references will increase shaping an annual cycle of media interest around the World Conferences.

We can see similar developments in the 290 references to climate change in *La Nación*. From almost the same figure in October (26), with Hugo Caligaris as special correspondent at the Project Sindycate event, references rise to 81 in December, almost double those published in *Clarín*, reflecting the inclusion of articles from other newspapers, op-eds and agency news concerning the summit. In January the failure in Copenhagen reduces the editorial impact (only 14 references compared to 27 in *Clarín*) but in the coming months many more details are published in *Clarín*, especially in March on the WWF campaign Earth Hour. For the remainder of the year the subject has a fairly regular presence in the daily agenda.

Sections and Supplements

First, we distinguish the daily pages from the weekly supplements since their productive routines (selection, prioritization, types of texts, thematic associations and authorship) vary. Consequently, of the 196 references to climate change that appear in the newspaper *Clarín* across the sample period, 159 are located in different sections of the daily pages (81.2 percent) while 37 appear in various specialized weekly supplements (18.8 percent).

Once this distinction is established, those news sections of the paper in which climate change appears can in turn be analyzed. It might be assumed that most reports will be published in the "Society" pages dedicated to the environment. However, in almost 50 percent of cases (78), the references appear in the *international news* section, headed "World" in *Clarín*. It is followed by "Society", which refers to climate change in 39 reports (24.5 percent).

In other news sections such as "Politics" and "Economics", the presence of climate change is much lower at 14 in the former and only two in the latter (8.8 and 1.3 percent, respectively). The opinion pages of the newspaper seem to publish reports about climate change at a remarkable rate: 16.4 percent.

The four supplements of *Clarín* in which climate change appears are the Rural Supplement, Architecture, Shows and Travel. However, it is Rural which has incorporated the topic into its own thematic agenda, with 26 references, 70 percent of the total reports. Entertainment and Architecture share matching numbers; in both supplements five items appear that reference climate change, representing 13.5 percent of total reports. The presence of the subject in Travel is residual: one short mention (2.7 percent).

In *La Nación*, 79 percent of the references (228) are found in the pages of the daily sections and in these, 44 percent of the references found are listed in the Foreign section while another large percentage is made up of those in General Information and Opinion with 14 percent each. The other important area is Science/Health, followed by Economics, with 8.3 percent of the reports, a figure much higher than in *Clarín*.

Three of the ten printed weekly supplements account for 63 percent of the references to climate change: Country, Focus and Community, with 23, 21 and 19 percent, respectively. Also prominent are the references in the supplement Foreign Trade (15 percent). The rest are sporadic: Magazine, Property, DNA Culture, Countries, Motoring and Technology, all with less than 5 percent of references on their pages, except the Magazine with 6.5 percent.

Climate Change as the Subject of Reporting

In *Clarín* the reports focused on climate change (the substantive reports dubbed Type I) account for 50 percent of the 486 articles, while those focused on more general environmental issues (Type II) account for 21.4 percent, and 28.6 percent is classified as Type III. In *La Nación*, 45.9 percent of reports focused on climate change compared to 30 percent of Type II and 24.1 of Type III (70 isolated references in texts) (Table 1).

The content analysis identifies 17 thematic categories: Agricultural Technology, Awareness-raising Activities, Food, Ecological Construction, Scientific Data, Environmental Deterioration, Economic Development, Climatic Phenomena, Impact on Biodiversity,

TABLE 1
Climate change news

	***Clarín* (*N* = 196)** *N* (%)	***La Nación* (*N* = 290)** *N* (%)
Type I (main topic)	98 (50)	133 (45.9)
Type II (one topic more)	42 (21.4)	87 (30)
Type III (only allusion)	56 (28.6)	70 (24.1)

TABLE 2
Thematic categories news Type I

	Clarín (*N* = 62) *N* (%)	*La Nación* (*N* = 84) *N* (%)
Awareness-raising Activities	3 (4.8)	15 (17.9)
Climatic Phenomena	2 (3.3)	7 (8.3)
Economic Development	1 (1.6)	5 (6.0)
Environmental Deterioration	5 (8.1)	0
Environmental Measures	4 (6.5)	3 (3.6)
Environmental Protests	2 (3.3)	2 (2.4)
Food	1 (1.6)	2 (2.4)
Health	1 (1.6)	1 (1.2)
International Politics	33 (53.2)	37 (44.0)
Scientific Data	9 (14.5)	11 (13.1)
Women	1 (1.6)	1 (1.2)

Energy Industry, Environmental Measures, Politics in Argentina, US Domestic Policy, Environmental Protests, International Politics, Health and Others.

In *Clarín*, International Politics is the thematic category that most frequently appears in news stories focused on climate change (Type I). Climate change appears to be linked to Climatic Phenomena in 28 Type II items. The second thematic frame is Argentinean Politics (20), while Economic Development and International Relations both have 12 units of analysis coded as Type II. In Type III pieces the determining thematic frame is International Politics with 73.2 percent of the total, an appreciable percentage that shows how climate change has become a matter incorporated into the policy agenda as one of the problems facing humanity.

Both newspapers highlight international negotiations in their climate change agenda, an issue that occupies 53.2 and 44.0 percent of Type I daily pages (opinion articles removed) reports in *Clarín* and *La Nación*, 62 and 84 reports, respectively (Table 2). *La Nación*'s second major theme is that of Awareness-raising Activities (17.9 percent), largely reflecting the extensive coverage of the Earth Hour campaign carried out by WWF, while in *Clarín* it is Scientific Data with 14.5 percent of the reports. This is smaller than that of *La Nación* despite Scientific Data occupying third place in the ranking of topics. Environmental Deterioration, the third of the subjects most published in *Clarín* (8.1 per cent), does not appear in *La Nación*, which instead selects more reports on climate phenomena explicitly related to climate change (8.3 percent of cases to *Clarín*'s 3.3 percent).

Generic Frames

Four frames are recognized to define and establish the terms of debate in news focused on climate change (Type I) in *Clarín* and *La Nación*: political conflict, risk, environmental sustainability and business opportunity.

The "conflict" frame is apparent in those reports that present a dichotomy or a confrontation between the protagonists of the media event. Nouns such as "controversial", "debate", or even "conflict" are used, as well as verbs such as "demand", "claim" or "defend", and belligerent terms like "war" or "battle".

TABLE 3
Generic frames news Type I

	Clarín (*N* = 74) *N* (%)	*La Nación* (*N* = 112) *N* (%)
Conflict	32 (43.2)	43 (38.4)
Risk	16 (21.7)	16 (14.3)
Sustainability	18 (24.4)	43 (38.4)
Opportunity	8 (10.8)	10 (8.9)

The "risk" frame is clear from the presentation of reports in typically catastrophic terms. The focus is placed on the worst predictions of the effects of climate change without any proposals for what action may be taken to prevent it.

The frame we have called "sustainability" appears in those reports that highlight the global environmental crisis and the need to change the situation. Developments in the fight against climate change in different sectors are reported using expert sources with an educative slant.

"Opportunity" shows climate change as a global situation that can be exploited to increase economic development, to do business.

The framework which prevails in the two leading newspapers in Argentina is that of "conflict", with 43.2 percent (*Clarín*) and 38.4 percent (*La Nación*), respectively, of Type I news (daily and supplements pages), in line with its prominence on the international politics agenda, as we have seen (Table 3). But in *La Nación* "sustainability" has the same percentage of published reports (38.4 percent), well above the figure for "risk" (14.3 percent), associated with scaremongering about the effects of climate change or the new data provided by climate scientists. In *Clarín* both frames offer similar percentages: 24.4 percent for "sustainability" and 21.7 for "risk". "Opportunity" with about 10 percent in both newspapers is the less common frame.

Conflict. This generic frame appears in almost half of the reports focused on climate change articles in *Clarín* and *La Nación*. Headlines such as "The Emerging Countries Fight Back and Demand More Efforts from the Rich" (Avignolo, *Clarín*, 11 December 2009) or "Battle Between Sceptics and Alarmists on the Internet" (*La Nación*, 10 December 2009) exemplify the "conflict" frame that surrounds items related to international negotiations.

In relation to the position of Argentina, special correspondent Gustavo Sierra (*Clarín*, 10 October 2009a) explained the positions of countries or groups of countries in international negotiations, concluding that "everyone is committed to somehow reducing carbon dioxide that their industries and people release into the air, but nobody is willing to implement measures that would involve the expenditure of billions of dollars if the 'other' does not do so first." Referring to Brazil, one of the actors to appear prominently in the press in Argentina and one of the biggest polluters of the world from deforestation and the burning of forests to create cattle grazing land, Sierra writes about the Brazilian bid to stop logging in the Amazon by 70 percent over the next 10 years, as long as the developed countries provide economic compensation for the losses this implies. "And here the issue of compensation enters the equation: how much money would the developed countries be willing to give to those countries which are still developing in order for them to stop polluting as the former have done until now in order to maintain growth?" (Sierra, *Clarín*, 10 October 2009a). In the same vein is "Ecuador, Ready to Renounce Petroleum if the World Pays" (Restivo, *Clarín*, 22 November 2009).

In an interview (Sierra, *Clarín*, 9 December 2009b), the then Director of Environmental Affairs, said that Argentina sought "greater ease of access to international funds", and spoke of the "environmental debt that countries which develop by polluting have with the rest".

In *Clarín*, we can see a perspective in the reports of the Copenhagen Summit by Gustavo Sierra, which is different to the European view; more so than in the statements of Argentinean politicians. In an article on December 10, 2009, for example, when a Danish draft was circulating at the summit, the correspondent of the Spanish daily newspaper *El País*, Rafael Méndez, wrote:

> The Danish draft stating that emissions should be peaking in 2020 raised the ire of the developing countries. The spokesman of these poor nations, the always-dramatic Sudanese Lumumba Stanislaus Kaw Di Aping, said that this would mean death for Africa. The chief negotiator of the European Commission, Artur Runge-Metzer, joked about his statements: "He lives in New York."

In contrast to the Spanish correspondent, Gustavo Sierra, the Argentine correspondent described these proposals as "unacceptable".

> The size of the industrial restructuring fund for poor countries is derisory...The disappointment is so great among the delegates from the least developed countries that in the corridors of the summit there is already talk of "carbon-century colonialism". They directly accuse the powerful of trying to force an agreement that would maintain the enormous inequalities that exist in today's world. (Sierra, *Clarín*, 10 December 2009c)

Di Aping is not dramatic:

> The president of the Group of 77, to which Argentina belongs, the Sudanese Lumumba Stanislaus Di-Aping, said that what those who wrote the document want is "to procure 60% of the globe's atmosphere for the richest 20%". And the Brazilian negotiator Sergio Serra said the document "does not include the position of 90% of the world." (Sierra, *Clarín*, 10 December 2009c)

These nuances are not in *La Nación* as most of their pieces about the summit, and in general their International pages, draw on information from international agencies.

Risk. In *La Nación* the "risk" frame is associated with scientific reports on the effects of climate change ("An impact that has already resulted in floods and droughts... As the twenty-first century progresses, the global picture could become increasingly bleak", García Oviedo, *La Nación*, 6 December 2009a) but especially the weather phenomena thematic frame: "2009 saw the worst weather in 50 years. This year there were natural disasters like floods, tornadoes, record temperatures and a severe deepening of drought" (Gallo, *La Nación*, 23 November 2009). All this under the heading of climate change.

The headline and the lead of the report "The Planet in Danger; last chance to stop climate change" (Sierra, *Clarín*, 10 October 2009a) illustrates the alarmist approach, dramatic, incorporating the news value of proximity. The opening of the text describes the impact of climate change in Argentina without specifying under what circumstances it would really take place:

> Half of the city of Buenos Aires under water. Most populations of the humid Pampa disappeared under huge salt lakes. The Northwest plunged into a permanent drought. "El Chocón" without water for years. Electricity generation by hydropower reduced to

less than half. What ensued was massive migration to the Andes of Patagonia in search of water. And this was only in Argentina. The rest of the world would be even worse. Half the world population would have nothing to drink. The other half, little to eat. And this is not an apocalyptic film.

Laura García Oviedo ("The Weather Would Result in Huge Losses in Latin America", *La Nación*, 17 December 2009b) wrote about the report submitted by the Economic Commission for Latin America and the Caribbean (ECLAC) in Copenhagen, which suggested an "immense" impact on economic activity, populations and ecosystems if urgent measures are not taken against climate change. Among the panellists was the Argentine Environment Secretary, Homero Bibiloni, who showed pictures of flooding and personalised the damage caused by the growing frequency and intensity of this kind of event. Bibiloni emphasized the catastrophic aspects, adding the human interest, and noted the Argentinean position, the same one adopted by the rest of the developed countries: industrialised nations have a "historical environmental debt to be paid", primarily with funding to implement various measures against global warming.

Sustainability. A clear example of the "sustainability" frame is the report of the journalist Sybila Camps "Uphill Battle to Capture the Chemicals That Harm the Planet" (*Clarín*, 29 November 2009): "With the environmental imbalance already in place, science steps up efforts to mitigate the effects. One of the ways is to catch and store gases that are causing the rising temperatures."

This detailed report published one Sunday, regardless of political events or publications of reports, with various national and international expert sources, starts from the idea outlined in the introduction that the exponential growth in fossil fuel use caused "an ecological imbalance on a global scale: global warming, caused by increased greenhouse gases (GHG) emitted by 80% of these fuels. Now it is up to science and technology to contribute to clearing up the mess."

In *La Nación* (17 December 2009), we find for example the report of Laura Corradini, correspondent in France, entitled "Faced With a Change of Lifestyle", which talks, among other things, about how humanity should adopt clean energy. The journalist writes that whatever the outcome of the climate summit in Copenhagen, "human beings will have to change their habits and invent new forms of consumption if they want to continue living, breathing and feeding on the planet."

Opportunity. "Opportunity" is the frame that does not appear in the *Clarín* daily and is only detected in a job-related report in the Economy section of *La Nación* (Raquel Saralegui, 16 October 2009, "Still, With Moderate Wind"), which outlines the potential workforce that would be created by a renewable energy industry. It should be pointed out that this frame is very important in Rural and Countryside supplements of the two newspapers, two supplements that define the image of Argentina in the world given that they reflect the interests and concerns of its main productive sector. In Rural, the reports focused on climate change are mainly related to the subjects of Economic Development and Agricultural Technology (27.3 percent each). However, most important is that in 63.6 percent of cases, the approach is that of "opportunity": "Facing the Challenges of the Future" (Persoglia, *Clarín*, 30 January 2010), "Argentina Should Play a Strategic Role" (Huergo, *Clarín*, 24 October 2009), "Carbon Is Good Business" (Neffren, *Clarín*, 20 August

2010) or "Change... Everything Changes" (Moreno, *Clarín*, 19 June 2010), in which, in a report on the impact of climate change on agriculture in Argentina, local specialist Graciela Magrin says there may be benefits to the country.

In *La Nación*, the Agricultural Technology thematic frame is still more prominent (88.9 percent of Type I pieces) and also the attribute "opportunity", 77.8 percent of cases. "The Countryside, the Challenge of Halting Climate Change" (Sanmartino, 19 December 2009b); "The World Offers Argentina an Opportunity" (Seifert, *La Nación*, 11 September 2009); "The Only Solution Is a Global Response", in a world where, according to Minister of Agriculture for New Zealand, food-producing countries have a future (Sanmartino, *La Nación*, 31 October 2009a).

Conclusions

Climate change is reported in Argentinean media in terms of the international political agenda and forms a cycle of annual media interest around the conferences on climate change hosted by the United Nations. Thus, information on climate change is an international issue for the press in Argentina and also for the Argentinean people as reflected in the Ibárometro survey published on December 30, 2009. According to this survey, the main desire of Argentineans for the world in 2010 was to end hunger and poverty (35.1 percent of respondents), and after that, the hope that solutions to pollution and global warming will be found (25.3 percent).

Our analysis reveal that in the Argentinean quality press climate change information is found mainly in the International section of newspapers instead of in Society (Science or more specifically the Environment) as might be anticipated. Therefore, the generic frame that appears most frequently is that of "conflict", which is associated with the various conflicting positions in international negotiations. The conflict does not appear in scientific debate about whether or not anthropogenic climate change exists. The antagonists are not scientists and sceptics since in no articles are the IPCC reports questioned; the Argentinean press does not present the level of scientific uncertainty that has been magnified by the media in other industrialized countries.

From the Argentinean perspective the confrontation is clearly between industrialized and developing countries. "They have a primary responsibility to reduce their emissions, an obligation to provide public funds to facilitate genuine mitigation," said the Argentinean Foreign Minister. While in the European press, "they" are the other industrialized countries that ratified the Kyoto Protocol and are not as involved as "we", the Europeans, in combating climate change; in contrast, in the press in Argentina "they" very widely means all developed countries. This goes at least for *Clarín*, through the chronicles of its special correspondent at the summit on climate change that shows this different approach to that of the European or North American. But this media construction is not found in *La Nación* as most of their pieces about the summit, and in the International pages in general, come from international agencies.

However, the "we" as Latin American countries does not appear as clearly defined as in Europe. Rarely is it mentioned that Argentina belongs to the Group of 77; nor is the role of Argentina well defined. Although the WWF Living Planet Report 2010 places Argentina among the 10 nations that make up more than 60 percent of the Earth's capacity to provide environmental services of global importance such as food production and carbon dioxide uptake, Argentina resembles more a spectator than an actor. The official position calls for a settling of the "environmental debt", but the fight against climate change is not a matter of priority on the political agenda in Argentina, at least seen from the perspective of the Argentinean quality press.

The oft-criticized emphasis on catastrophe associated with environmental journalism appears in about 25 percent of the articles and nearly always relates to scientific reports or in relation to extreme weather events. More news falls under "sustainability", which is predominant in environmental reporting that promotes "eco-literacy", environmental education through information on environmental issues released by the media (Fernández Parrat, 2006, p. 52). This is not about taking sides but about giving the environment the place and depth that it deserves in media coverage. In this regard, Argentinean journalists such as Sergio Elguezabal seem to have a clear role in promoting environmental values: "Besides the political decision to undertake an integrated approach to the issue, more rigorous, less grumpy journalism with continuing commitment is vital" (*La Nación*, 9 November 2009).

There has to be better communication, says the editor of the Science/Health section of *La Nación*, Nora Bär, because according to a survey in the United States, four in ten did not know what a fossil fuel is, and six out of ten were equally in the dark about renewable energy (24 February 2010).

Hugo Caligaris (*La Nación*, 13 October 2009) explained very clearly what many expect of journalists in this crisis, "Journalists are the only ones capable of generating a strong current of opinion so as to force the rulers to correct course. In part, they already have: The subject is on the table there is more awareness than a few years ago. But we have to fight more vigorously."

ACKNOWLEDGEMENTS

The author would like to thank the grant Santander UCH CEU Mobility Scholarship 2010 at the Universidad Austral (Buenos Aires, Argentina).

NOTE

1. In order to assess intercoder reliability, two external encoders who had not participated in the initial research encoding carried out the coding process (Krippendorff, 2004; Neuendorf, 2002). Fifty reports were selected randomly (26.8 percent of the total units Type I) and were distributed to the two analysts (the same 50 for each). They both gathered the data according to the criteria established in the codebook. The data generated by the encoders were used to create a double encoding data matrix using SPSS V.14.0 Windows. Two criteria were used for the analyses: percentage of observed agreement and Scout coefficient Pi ($-1/+1$). As a result, the basic variables "section" and "themes" yielded a reliability of 0.97 and 0.91, respectively. Regarding the variable "frames", this was an average of 0.69 and a percentage agreement of 92, indicating that the reliability between the encoders was proper, taking into account the existent standards.

REFERENCES

ANDERSON, ALISON (1997) *Media, Culture and Environment*, London: UCL Press.

ANDERSON, ALISON (2009) "Media, Politics and Climate Change: towards a new research agenda", *Sociology Compass* 3(2), pp. 166–82.

AVIGNOLO, Mª LAURA (2009) "Los países emergentes contraatacan y exigen más esfuerzos a los ricos", *Clarín*, 11 December.

BACHETTA, VÍCTOR L. (2002) "Perfil del periodista ambiental", *Sala de Prensa* 42(2), http://www.saladeprensa.org/art340.htm, accessed 29 September 2008.

BÄR, NORA (2010) "Hay que comunicar mejor", *La Nación*, 24 February.

BERELSON, BERNARD (1952) *Content Analysis in Communication Research*, New York: The Free Press.

BOYKOFF, MAXWELL T. and BOYKOFF, JULES M. (2004) "Bias as Balance: global warming and the US prestige press", *Global Environmental Change* 14, pp. 125–36.

BOYKOFF, MAXWELL T. and ROBERTS, J. TIMMONS (2007) *Media Coverage of Climate Change: current trends, strengths, weaknesses*, United Nations Development Report, Occasional Paper, 2007/3.

BROSSARD, DOMINIQUE, SHANAHAN, JAMES and MCCOMAS, KATHERINE (2004) "Are Issue-cycles Culturally Constructed? A comparison of French and American coverage of global climate change", *Mass Communication and Society* 7(3), pp. 359–77.

CALIGARIS, HUGO (2009) "Llamada mundial para salvar el planeta", *La Nación*, 13 October.

CAMPS, SIBILA (2009) "Dura batalla para capturar las sustancias que dañan el planeta", *Clarín*, 29 November.

CAMPS, SIBILA (2010) "La capa de ozono ya no se debilita", *Clarín*, 30 June.

CARVALHO, ANABELA (2008) "Communicating Climate Change: discourses, mediations and perceptions", Centro de Estudios de Comunicación y Sociedad, Universidad de Minho, pp. 126–56.

CARVALHO, ANABELA and BURGESS, JACQUELIN (2005) "Cultural Circuits of Climate Change in the UK Broadsheet Newspapers", *Risk Analysis* 25, pp. 1457–70.

CORRADINI, LAURA (2009) "Ante un cambio del modo de vida", *La Nación*, 17 December.

DE VREESE, CLAES H. (2003) *Framing Europe. Television news and European integration*, Amsterdam: Aksant.

DÍAZ NOSTY, BERNARDO (2009) "Cambio climático, consenso científico y construcción mediática. Los paradigmas de la comunicación para la sostenibilidad", *Revista Latina de Comunicación Social* 64, pp. 99–119.

ELGUEZABAL, SERGIO (2009) "El mundo acabará mañana", *La Nación*, 9 November.

ENTMAN, ROBERT (1993) "Framing: toward clarification of a fractured paradigm", *Journal of Communication* 43(4), pp. 51–8.

FERNÁNDEZ PARRAT, SONIA (2006) *Medios de comunicación y Medio Ambiente*, Madrid: Fragua.

FÜRSICH, ELFRIEDE (2009) "In Defense of Textual Analysis", *Journalism Studies* 10(2), pp. 238–52.

GALLO, DANIEL (2009) "En 2009 hubo el peor clima en 50 años", *La Nación*, 23 November.

GARCÍA OVIEDO, LAURA (2009a) "Un impacto que ya se traduce en inundaciones y sequías", *La Nación*, 6 December.

GARCÍA OVIEDO, LAURA (2009b) "El clima dejaría cuantiosas pérdidas en América latina", *La Nación*, 17 December.

GITLIN, TODD (1980) *The Whole World Is Watching: mass media in the making and unmaking of the new left*, Berkeley: University of California Press.

GORDON, JOYE, DEINES, TINA and HAVICE, JACQUELINE (2010) "Global Warming Coverage in the Media: trends in a Mexico City newspaper", *Science Communication* 32(2), pp. 143–70.

HUERGO, HÉCTOR A. (2009) "Argentina debe jugar un rol estratégico", *Clarín*, 24 October.

IGARTUA, JUAN J. (2006) *Métodos cuantitativos de investigación en comunicación*, Barcelona: Bosch.

IGARTUA, JUAN J. and HUMANES, M. LUISA (2004) *Teoría e investigación en comunicación social*, Madrid: Síntesis.

KIM, SEI-HILL, SCHEUFELE, DIETRAM A. and SHANAHAN, JAMES (2002) "Think About It This Way: attribute agenda-setting function of the press and the public's evaluation of a local issue", *Jounalism & Mass Communication Quarterly* 79, pp. 7–25.

KOTHARI, AMMINA (2010) "The Framing of the Darfur Conflict in *The New York Times*: 2003–2006", *Journalism Studies* 11(2), pp. 209–24.

KRIPPENDORFF, KLAUS (2004) *Content Analysis: an introduction to its methodology*, 2nd edn, Thousand Oaks, CA: Sage.

LA NACIÓN (2009) "Batalla entre escépticos y alarmistas en Internet", 10 December.

LADLE, RICHARD, JEPSON, PAUL and WHITTAKER, ROBERT J. (2005) "Scientists and the Media: the struggle for legitimacy in climate change and conservation science", *Interdisciplinary Science Reviews* 30(3), pp. 231–40.

MARTÍN PÉREZ, IDAFE (2010) "Con el fantasma de una crisis que no cesa, España presidirá la UE", *Clarín*, 1 January.

MCCOMBS, MAXWELL and SHAW, DONALD L. (1972) "The agenda-setting function of the mass media", *Public Opinion Quaterly* 36, pp. 176–87.

MCCOMBS, MAXWELL, LLAMAS, JUAN PABLO, LOPEZ-ESCOBAR, ESTEBAN and REY, FERNANDO (1997) "Candidate Images in Spanish Elections: second-level agenda-setting effect", *Journalism & Mass Communication Quarterly* 74, pp. 703–17.

MÉNDEZ, RAFAEL (2009) "EE UU excluye a China de las ayudas contra el cambio climático", *El País*, 10 December.

MORENO, ELINA (2010) "Cambia . . . todo cambia", *Clarín*, 19 June.

NEFFREN, GASTÓN (2010) "Es carbono es un buen negocio", *Clarín*, 20 August.

NELKIN, DOROTHY (1990) *La ciencia en el escaparate*, Madrid: Fundesco.

NEUENDORF, KIMBERLY A. (2002) *The Content Analysis Guidebook*, Thousand Oaks, CA: Sage.

NOVO, MARÍA (2003) *La educación ambiental. Bases éticas, conceptuales y metodológicas*, Madrid: Editorial Universitas.

OLAUSSON, ULRIKA (2009) "Global Warming–Global Responsibility? Media frames of collective action and scientific certainty", *Public Understanding Science* 18, pp. 421–36.

PERSOGLIA, SERGIO (2010) "Los desafíos que llegan del futuro", *Clarín*, 30 January.

PIÑUEL, J. LUIS (2002) "Epistemología, Metodología y Técnicas del Análisis de Contenido", *Estudios de Sociolingüística* 1, pp. 1–42.

PIZARRO, MARCELA (2008 *El "mundo" de la prensa argentina: ¿Qué es noticia internacional Para La Nación y Clarín?*, Buenos Aires: La Crujía.

RESTIVO, NÉSTOR (2009) "Ecuador, dispuesto a resignar petróleo si el mundo lo paga", *Clarín*, 22 November.

SÁDABA, TERESA (2001) "Origen, aplicación y límites de la 'teoría del encuadre' (framing) en comunicación", *Comunicación y Sociedad* 2, pp. 143–75.

SANMARTINO, FÉLIX (2009a) "La única solución es una respuesta global", *La Nación*, 31 October.

SANMARTINO, FÉLIX (2009b) "El campo, ante el reto de detener el cambio climático", *La Nación*, 19 December.

SARALEGUI, RAQUEL (2009) "Todavía, con viento moderado", *La Nación*, 16 October.

SCHÄFER, MIKE S., IVANOVA, ANA and SCHMIDT, ANDREAS (2011) "Media Attention for Climate Change", KlimaCampus, University of Hamburg, http://www.klimacampus.de/macc.html, accessed 28 October 2011.

SCHEUFELE, DIETRAM A. (1999) "Framing as a Theory of Media Effects", *Journal of Communication* 49(1), pp. 103–22.

SEIFERT, ROBERTO (2009) "El mundo ofrece una oportunidad para la Argentina", *La Nación*, 11 September.

SEMETKO, HOLLI and VALKENBURG, PATTI M. (2000) "Framing European Politics: analysis of press and TV news", *Journal of Communication* 50(1), pp. 93–109.

SIERRA, GUSTAVO (2009a) "El planeta en peligro", *Clarín*, 10 October.

SIERRA, GUSTAVO (2009b) "Diseñar nuestros proyectos", *Clarín*, 9 December.

SIERRA, GUSTAVO (2009c) "Países ricos y pobres chocan por el costo de reducir la contaminación", *Clarín*, 10 December.

TANKARD, JAMES W. JNR (2001) "The Empirical Approach to the Study of Media Framing", in: Stephen D. Reese, Oscar H. Gandy and August E. Grant (Eds), *Framing Public Life: perspectives on media and our understanding of the social world*, Mahwah, NJ: Lawrence Erlbaum Associates, pp. 95–106.

VICENTE, MIGUEL and LÓPEZ, PABLO (2009) "Resultados de la investigación en comunicación sobre framing: sólido avance y arranque de la especialidad en España", *ZER Revista de Estudios en Comunicación* 26, pp. 13–34.

WEINGART, PETER, ENGELS, ANITA and PANSGRAU, PETRA (2000) "Risks of Communication: discourses on climate change in science, politics, and the mass media", *Public Understanding of Science* 9(3), pp. 261–83.

WU, DENIS and COLEMAN, RENITA (2009) "Advancing Agenda-setting Theory: comparative strength and new contingent conditions of the two-levels of agenda-setting effects", *Journalism & Mass Communication Quarterly* 86, pp. 775–89.

PARTICIPATORY POLITICS, ENVIRONMENTAL JOURNALISM AND NEWSPAPER CAMPAIGNS

Anita Howarth

This article explores the extent to which approaches to participatory politics might offer a more useful alternative to understanding the role of environmental journalism in a society where the old certainties have collapsed, only to be replaced by acute uncertainty. This uncertainty not only generates acute public anxiety about risks, it has also undermined confidence in the validity of long-standing premises about the ideal role of the media in society and journalistic professionalism. The consequence, this article argues, is that aspirations of objective reportage are outdated and ill-equipped to deal with many of the new risk stories environmental journalism covers. It is not a redrawing of boundaries that is needed but a wholesale relocation of our frameworks into approaches better suited to the socio-political conditions and uncertainties of late modernity. The exploration of participatory approaches is an attempt to suggest one way this might be done.

Introduction

It only makes sense to redraw the boundaries between journalism's established demarcations of objective versus polemical journalism, and reportage versus commentary if the premises underpinning these concerning the ideal role of the media in society and journalistic professionalism are considered valid. This article argues that this is not the case: that aspirations of objective reportage are outdated and ill-equipped to deal with many of the new risk stories environmental journalism covers. It is not a redrawing of boundaries that is needed but a wholesale relocation of our frameworks into approaches better suited to the socio-political conditions of late modernity.

The roots of current Anglo-American journalistic ideals and practices lie in the emergence of classic liberalism during the Age of Revolution of the eighteenth and nineteenth centuries where, on the one hand, there was concern about the power of the press to create political stability or foment civil unrest. On the other hand, there was considerable enthusiasm and optimism about the prospect of societal "progress" from the new methods and inventions of the scientific-industrial revolution. What emerged were ideas about the democratic role of the press as a watchdog against the abuses of corrupt politicians whose excesses risked inciting civil unrest, the provider of accurate information necessary for an informed public and a forum for critical-rational debates that provided a necessary safety valve for tensions in society and communicating the wishes of the public to government (see Scammell and Semetko, 2000). The professional mechanisms for delivering these—neutrality, "objectivity" and commitment to factual accuracy—came, over the next 250 years, to be seen as the defining qualities of the "professionalism" necessary to meet the "higher goals" and "duties" to democracy (Scammell and Semetko, 2000). The supposed output of this professionalism was an "objective journalism 'of record' that [drew] on authoritative sources to provide reliable information of matters of significance" (McQuail, 2000, p. 32). Conversely, certain types of polemical writing in certain places and at certain times could then be criticized and dismissed as an unprofessional debasing of these ideals and potentially irresponsible because of the potential power of the press to incite public sentiment and to destabilize the political system (see Scammell and Semetko, 2000). Thus, the preoccupation with "objective" reporting is a historical one rooted in notions of what is needed for a stable society.

As Scammell and Semetko note, what is "curious" is that this classic view, rooted as it is in historical assumptions about political stability and societal progress, and the ideal roles of the press and the conduct needed to ensure this, has continued to "dominate so much thinking about the modern media" despite substantial societal changes since its origins and despite the rejection of classic liberalism by political theorists as "outdated" (2000, p. xii). Notwithstanding these critiques, dominant strands in journalism studies have remained pre-occupied with questions of how, why and with what implications current journalism practice falls short of the ideal (see, for instance, the debates on the "crisis of communication"; Blumler and Gurevitch, 1995).

Environmental journalism, however, exposes the inadequacies of these types of evaluations and of traditional polarizations between objective and polemical writing, reportage and commentary. It is able to do this because much of its subject material is concerned with new risks—oil spills, genetically modified (GM) food and crops, nuclear power, climate change—where classic demarcations in journalism collapse, not because of unprofessional practices but because of the nature of the debates. These debates are characterized by scientific disputes over evidence, knowledge and facts so any notion of "objectivity" in science and the "objective" reporting of this by journalists becomes oxymoronic.

The challenge for journalists is that disputes about knowledge in conditions where there is no clear evidence of harm and no clear evidence of safety are abstract, remote from the public and couched in scientific or academic jargon that renders them inaccessible to a lay audience and to many journalists (see Hargreaves and Lewis, 2003; Kitzinger and Reilly, 1997). However, when basic essentials—food, air, water—are implicated, otherwise esoteric debates become directly relevant to the public and so legitimate subject matter for journalists. The challenge is how to make the esoteric accessible to a lay readership and here visual images of dying fish (Beck, 1992 [1986]), metaphorical spectres of disaster such as "Frankenstein foods", or parallels with other scares can play a crucial role (see Howarth, forthcoming). When journalists are able to make these connections the media can cast the "spotlight" on new risks and their relevance for readers (Beck, 1992 [1986]). When the risks impinge on life-sustaining activities, implicate individual decisions on what to eat or drink, and are located in the wider context of a history of government mismanagement of food scares (Howarth, forthcoming), then strong public emotions such as fear and anxiety become reasonable rather than irrational responses to perceived threats. However, the classic liberal emphasis on rational debate, objective reporting and factual accuracy undermines the capacity of journalists to articulate these without being accused of sensationalism.

It was in this context that four national newspapers in Britain launched campaigns on GM food with the agenda of "revealing the hidden facts in food", educating the public about these and advocating policy change. The response of the then Prime Minister Tony Blair to these campaigns was to denounce them as "sensationalist", fictionalized campaigns of "misinformation" (Blair, 1999). When he resigned eight years later Blair (2007) claimed that journalistic "accuracy" had become "second to impact"; that "rather than just 'report' news", journalists were now providing "interpretation"; and that the old boundaries between news and commentary, fact and opinion had blurred. In so doing, Blair drew on classic liberal assumptions about the ideal role of the media in society, the professional practices that are said to underpin this and the nature of the media–state relationship.

However, as argued above, these assumptions make little sense when the subject matter of environmental journalism is concerned with new risks. Aspirations of journalistic "objectivity", accuracy and neutral reportage infused from an era of authoritative science are incompatible with conditions of radical uncertainty where knowledge itself is disputed, the facts are contested, debates are highly polarized, policy elites are not trusted to act in the public interest and new moral discourses are emerging about consumer rights and agency.

This article explores the extent to which approaches to participatory politics might offer a more useful alternative to understanding the role of environmental journalism in a society where the old certainties have collapsed, only to be replaced by acute uncertainty. It then applies this to a discourse analysis of the editorials on GM food in four newspapers that campaigned against it and explores the extent to which assumptions about participatory politics might be more useful than the classic liberal premises about the role of the media in acting as a watchdog, providing "quality" of information and facilitating debate.

Classic Liberalism, the Age of Revolution and Science

The classic liberal view of democracy and the role of the media emerged in response to particular historical conditions and fears. In eighteenth-century America, it appeared in the argument of James Madison, writing in the midst of fears that civil war could tear apart the new republic (see Kaminsky, 2006), for the constitutional protection of freedom of expression as a necessary "safety valve" for the expression of diverse views and hence a mechanism for political stability. Across the Atlantic it appeared in the writings of J. S. Mill amid anxiety that the wave of revolutions sweeping Europe could spread to Britain if the media were not free to expose and so check the abuses of corrupt politicians before they could incite civil unrest. The free press was therefore seen as an essential outlet for pent up frustrations and the confining of the "enormities of the revolution" (Mill, cited in Hargreaves, 2003, p. 44).

However, revolution was not only political but also scientific and technological. There was much enthusiasm about what innovations might mean for societal "progress" and this infused notions of an emerging profession of journalism. Thus, Mill argued that free expression "is fundamental not only to political freedom, but to society's ability to gather knowledge empirically and so to progress scientifically and in other domains" (cited in Hargreaves, 2003, p. 44). Thus from an early stage in the development of the core tenets of classic liberalism, fundamental links were made between press freedom, public opinion and mechanisms for political stability and progress.

Enthusiasm for the scientific revolution also informed notions about the kinds of news-gathering methods deemed most appropriate. Here writers such as Mill drew on two separate strands of thinking: the valuing of scientific method—the application of the rational mind to the dispassionate observing of a phenomenon in order to accumulate objective, factual and reliable knowledge—and fear of those forms of speech which could incite protests, civil unrest and cause harm. The ability of scientific methods to deliver unitary, objective and authoritative science (Locke, 2001; Yearley, 1991) and deliver a dispassionate, impartial account of what was observed were therefore seen by Mill and others as also capable of delivering objective journalism that served the higher ideals of stable democracy and political order. Thus the principles of scientific method were distilled into notions of what constituted professional conduct, practice and its contribution to society (McQuail, 2000).

Notwithstanding the persistence of discourses of "objectivity" in debates about journalistic standards, the concept has been poorly defined (McQuail, 2000; Schudson, 2001) largely because it has been divorced from these historical origins and partly because of the desire of journalists to deliver believable accounts of what happened (see Mindich, 2000; Schudson, 2001). Objectivity, when applied to journalism, was taken to mean a pseudo-rational style of writing which claimed to be an unemotional, impartial and accurate reporting of the facts (Mindich, 2000, p. 8). In providing this, journalists would not only provide the information the public needed to make political decisions on who to vote for, they would also be able to stand as watchdogs, offering a disinterested, dispassionate and independent scrutiny of the abuses of those in political power. This did not mean opinion pieces were excluded from newspapers, but they were carefully located within a layout that ensured clear distinctions between objective journalism and polemical pieces, reportage and editorial opinion (see Schudson, 1978). Even then any overt advocacy tends to provoke fierce debate within journalistic circles (see Neuzil, 2008).

In Britain, these divisions are not absolutes. During elections national newspapers openly declare support for one or other party and then claim they influenced the outcome (see Thomas, 2005). More important, however, for this article is a century-old tradition of a "campaigning press" (Birks, 2009; Greenslade, 2004) and the construction of newspaper identities around what is openly an endeavour in polemics and advocacy (see Conboy, 2006). However, this historically rooted co-existence of objective and polemical roles in the same newspapers has remained one of the largely unexplored paradoxes in British newspapers primarily because of the persistence of classical liberal paradigms.[1] When viewed through the prism of these ideals such campaigns signify to critics a debasing of the higher ideals of "objective" journalism and its role in contributing to "rational" debate. When viewed through the prism of new risk-type environmental stories such overt partiality can be justified through an over-arching public interest evoked by the "potentially politically catastrophic" consequences of the advances of late modernity (Beck, 1992 [1986]). An alternative framework is needed to address the paradox of co-existing objective and polemical ideals.

However, before exploring one such possibility it is worth reflecting on the persistence of classic liberalism and its associated notions of journalistic objectivity. Notwithstanding the discrediting of objective and impartial journalism as a "myth" (Schudson, 1978) with the "lynching" of African-Americans in the 1890s (Mindich, 2000) and media coverage of the McCarthy era (Cunningham, 2003), it has persisted. These cases may have discredited the notion but they failed to undermine the broad valuing of the "scientific method" and the ideals associated with it. Schudson suggests these persisted because journalists continued to believe them; they "wanted to, needed to, were forced by ordinary human aspiration to seek escape from their own deep convictions of doubt and drift" (1978, p. 21). However, the pervasive, radical uncertainty of new risks (see Beck, 1992 [1986]) challenges the ability of journalists to "escape" when dealing with topics such as climate change, GM food and crops, etc. If objective, authoritative science is not possible when the evidence is contested, facts are disputed and the method of knowledge construction are the subject of disagreements, it then becomes difficult to see how it is possible to have objective, impartial and factually accurate reporting of these. In this context, the persistent adherence in journalism studies to the tenets of objectivity, impartiality and factual accuracy within classic liberalism needs to be questioned, not least because of its "taking for granted" of a narrow view of the roles of the media then using this to "provide the premises" for "virtually all avenues of inquiry" (Scammell and Semetko, 2000, p. xii), including questions of professional output.

There is a variety of twentieth-century theories of democracy that offers alternatives and there is an urgent need for journalism studies to explore these. The caveat, though, is that twentieth-century alternatives to classic liberalism "rarely focus on the mass media", so it is necessary to "tease out inferences" for the media and for journalism (Scammell and Semetko, 2000).

The Alternative Offered by Participatory Approaches to Politics

Participatory approaches to politics and democratic theory tend to ignore the media (Scammell and Semetko, 2000, p. xxxvii); however, it is possible to adapt some of their principles (see Teorell, 2006) to outline the role and activities the traditional media may play in such approaches.

First, participatory approaches argue that political apathy in mature democracies is a hallmark of public disillusionment, alienation and disengagement with the conduct and relevance of politics for individual citizens (see Pateman, 1970; Teorell, 2006). The revival of democracy, they argue, lies in the valuing of political participation as an "end in itself" and not merely as a means to an end, for instance with the electing of new leaders, as classic liberal approaches have tended to assume (Scammell and Semetko, 2000; Teorell, 2006). This facilitates a more expansive notion of participatory activities that legitimizes many diverse forms of environmental protest. It also means repositioning the media, not as the dispassionate and impartial observers of classic liberalism, but as key, active and engaged participants in political debate with the potential to enhance public participation through educating, informing, etc. The media can therefore play a crucial role in realizing the value of participation as an "end" in itself; so the key questions then are not the quality of media participation but the existence of the participation, the potential it has for expanding the participation of others and the means by which they are able to do this. Rather than fearing the potential for the media to incite resistance and civic unrest, as classic liberalism tends to do, participatory approaches value this potential as a legitimate "end" in itself and as a means of revitalizing public participation, for instance in environmental protests in mature democracies characterized by political apathy.

Second, participatory theorists such as Pateman (1970) challenge assumptions that individuals are "free and equal" in liberal democracies and argue that the formal or legal existence of certain rights is of limited value if they do not translate into practical choices in everyday life (see also Held, 2006). This means that the site of struggle therefore is the lived experience of here and now, not possible choices at the next election. It also means that claims about existing relationships need to be examined in order to assess whether freedoms are operating in practice.

Third, Teorell (2006) includes discussion as a form of participatory activity. Liberal and deliberative models of democracy (see Dryzek et al., 2006; Habermas, 1989 [1962], 1992) have long afforded the media a key role in the public sphere where diverse viewpoints are articulated in critical-rational debates, collated into a consensus about collective opinion and then transmitted to governments. The problem with this framework is that emotive discourses are not only anathema to critical-rationalists but also the antithesis of apathy and a key component in political engagement and mobilization. The denial of the legitimacy of such discourses within the public sphere signifies an "emotional deficit" in these approaches (see Richards, 2004), but their inclusion provides politicians with the means to discredit certain types of newspaper coverage as "irrational" or "sensationalist" (see Blair, 2007). Participatory approaches do not de-legitimatize emotion in politics, so it becomes possible to interweave both feelings and reason within what may be considered legitimate or credible public discussion. The other problem with liberal and deliberative approaches to discussion is their concern with the formation of a consensus in which certain views can be excluded, marginalized or silenced within claims of a "majoritarian" consensus. The assumption in participatory approaches is that there will be a multiplicity of perspectives, articulated in multiple forums; even when one side loses the policy debate the final decision can still have legitimacy if the different sides of an argument have been heard. So it is not the type of discourse, consensus formation or the influence on the decision that confers legitimacy but participation and the arrival at a decision through the consideration of different views that matters (Teorell, 2006 p. 797).

The fourth type of participatory activity, that of "influencing attempts" (Teorell, 2006, p. 788), is arguably the most controversial when applied to journalism and offers one way of addressing the objectivity–polemics paradox in Britain's campaigning newspapers. Self-labelled campaigns are deliberate, explicit attempts to change the direction of government policy. That is, they signify the intentional shift in a newspaper from assumptions of "impartial" or "objective reporting" to advocacy and to mobilize the public (see Howarth, forthcoming). Locating these forms of newspaper engagement within participatory approaches means that rather than campaigns being dismissed as deviations from professional standards or remaining silent on what they say about journalism it becomes possible to consider them as legitimate forms of participatory politics. In which case, partiality becomes acceptable.

The analysis in this article applies these four principles to newspaper engagement with GM food and crops—a new risk domain in which government assumptions about objectivity and certainty were increasingly challenged.

New Risks, Food Scares and Distrust of Government—Challenges to Objectivity and Certainty

The central contention here is that classic liberal assumptions about the press and objective journalism are outdated in late modernity. The reporting of GM food and crops illustrates this well.

Societal disputes in Britain over GM food and crops came to embody typical new risk debates. New risks are qualitatively different from natural disasters or controllable man-made risks in that they are manufactured through the advanced technological processes of late modernity, and their potential consequences are unknown and potentially unlimited (Beck, 1986 [1992], 1999; Giddens, 1998). However, the disputes over new risks are essentially over knowledge, what counts as evidence and what method may be most appropriate for determining evidence, whether there is evidence of harm, and whether, in the absence of evidence of harm, it can be assumed the new technologies are safe. In other words, scientific uncertainty and conflict exist at every level of these disputes. The idea that certainty can be derived from objective, authoritative science and the accumulation of more knowledge becomes meaningless when even the methods of achieving the knowledge are disputed by scientists (see Pusztai, 2002).

The potential for acute public uncertainty and heightened anxiety lies in the pervasiveness and proximity of the threats that new risks such as GM food and crops are seen to pose to everyday life-sustaining activities through the contamination of air, food, water, etc. (Beck, 2003; Giddens, 1998). However, they are also seen as undetectable to the naked eye, so ordinary people are powerless to take evasive action unless GM products are labelled as such. Governments, rather than protecting the public, are seen to be complicit through policies that encourage the expansion of the new technologies but are unable to limit the possible adverse impacts because knowledge is too limited (Beck, 2003; Giddens, 1998).

Beck (1992 [1986]) argues that the media have a central role to play here in "spotlighting" the risks, exposing the scientific disputes over evidence to the public gaze and challenging the decisions of government to subject ordinary people to risks they cannot manage for want of adequate knowledge (Beck, 1992 [1986]; see also Cottle, 1998). That is, the role of the media is to expose fallacious claims of objective, authoritative science, but they cannot do so through claims of counter-objectivity because the very nature of knowledge and the methods of acquiring it are in dispute. Furthermore, the news media are "ill designed" to cover stories about such environmental risks and debates partly because the disputes are about highly abstract, technical, futuristic risks (Kitzinger and Reilly, 1997, p. 320) and partly because of a pre-occupation with objectivity, factual accuracy and impartiality—the legacy of classic liberal assumptions about journalistic professionalism. What made GM food and crops particularly difficult for journalists was the absence of deaths, of incontrovertible evidence of harm or any visual evidence of contamination. While news routines usually rely on the concrete, the actual and the visual, GM food and crops presented journalists with debates about the abstract, hypothetical and invisible.

The first two principles of a participatory media—widening engagement and highlighting issues of everyday relevance and disempowerment—are particularly useful here. As newspaper engagement with GM food issues intensified, retailers claimed there had been a marked fall in sales of GM foods that culminated in a *de facto* consumer boycott (Howarth, 2010). School and work canteens banned GM food, GM crop fields were trashed, individual consumers staged protests in shops and a series of legal suits by individual farmers and consumers were brought against the big agribusiness companies for contamination of their farmland. Within weeks of the launch of campaigns against policy by four newspapers a *de facto* retailer boycott joined the consumer one. A participatory approach could argue that the issue here was not whether these newspaper responses were rational or whether direct causal effects linking media, specific behavioural actions and policy change can be proved, but the ability for increased newspaper engagement to mobilize public engagement and their willingness to become active, partisan and overt participants in policy debates.

The second principle was most clearly evident in that the new risks implicated food and everyday staples such as bread that everyone was consuming. One of the key issues was the extent to which consumers were able to choose whether or not to consume GM food. The government claimed it had introduced mandatory labelling (Blair, 1999); consumer and environmental non-governmental organizations argued that this applied to a very small proportion, that 60 per cent of the food on supermarket shelves containing GM ingredients were unlabelled (Austin and Lo, 1999). This was to become a core plank in newspaper opposition to GM policy, but in order to mount such an opposition, facilitate discussion and attempt to influence policy, journalists needed to overcome the abstract, hypothetical nature of the subject matter and address the collapse of the argument for scientific objectivity.

The Analysis of Campaigns Against GM Food and Crops

Given these professional challenges it is hardly surprising that initial newspaper coverage of GM food and crops was sporadic, fragmentary and contradictory despite a decade of food scares including BSE/CJD (Howarth, 2010). It was only when direct parallels were drawn between BSE/CJD and the cultural/metaphoric discourses of "Frankenstein foods" and human "guinea pigs" were routinely used that journalists were able to engage more fully with the debates. The following analysis explores how these links were made, entrenched and challenged by four critical interventions—Prince Charles and Professor Árpád Pusztai in mid-1998, then the launch of the campaigns by the end of February 1999 and Blair's counter-attack on these. This rare direct criticism of the newspapers by a prime minister was based on classic liberal assumptions and ideals about objective, factually accurate reporting. The newspapers' justifications and defence of these campaigns articulated the fallacies in these assumptions. Presented with this polarization, the analyst is left with a choice—either the newspaper campaigns were an aberration, a deviation from professional standards of objective, impartial journalism, or these notions are inadequate for addressing what is the professional responsibility of journalism when dealing with new risks and escalating public resistance to these. This article subscribes to the latter perspective and argues that an alternative view of the role of the media is needed; it here draws on participatory approaches to democracy, in particular newspapers' roles in "influencing attempts"—of public and government—and in "discussion".

This article defines newspaper campaigns as: (1) the conscious editorial decision by a particular newspaper directly and explicitly to intervene in a policy debate with the stated intention of influencing policy change or mobilizing public opposition to policy; (2) the self-labelling of this form of intervention as a "campaign" and the justification of this departure from what critics would claim was the role of "objective" or "impartial" purveyors of information (Howarth, forthcoming). The discourse analysis that follows focuses on the four newspapers that launched such campaigns. Critics could argue that this is self-defining; however, the counter-argument is that the approach taken in this article is informed by the premises of participatory politics and so privileges a focus on campaigns as a particular form of media participation. Furthermore, the four titles spanned the ideological and demographic reach of British newspapers from the centre-right *Daily Mail* and the pro-Blair *Express*, which dominated the growing and politically significant mid-market (Greenslade, 2004); the *Independent on Sunday*, which was pro-environment and pro-business on regulatory issues while also claiming to be "independent" of party and commercial interference; and the *Mirror*, which was a working-class, pro-Labour title. (For more on these ideological predispositions, see Greenslade, 2004, and evidence given by editors to various parliamentary select committees.)

Even while restricting the focus to the four campaign titles, a Nexis news search between 1998 and 2000 delivered over 4000 articles, so the discourse analysis was restricted to editorials. Editorials are widely accepted as a legitimate forum for the routine and explicit articulations of "the opinion of the newspaper" on an issue (Richardson and Lancendorfer, 2004, pp. 75, 80), sites where newspapers "make use of their right to present themselves as autonomous actors" (Eilders, 2002, p. 26; Firmstone, 2008, pp. 212–13). This opinion explicitly interprets the issues of the day in terms of the particular ideology, values and positioning of the newspaper (Van Dijk, 1996). It also provides a forum in which editors can directly address particular audiences, such as policy elites, and defend any criticisms of their coverage (Henry and Tator, 2002; Izadi, 1997, p. 141). Editorials are thus the one place where the newspaper's constructions of its participatory roles are likely to be most explicitly articulated as well as justified. In articulating the newspaper's position on an issue, editorials also serve as an organizing fulcrum around which the tone and content of news stories will be structured. While alternative viewpoints may well be given space, the dominant position across all the pages should broadly reflect the editorial policy articulated in these columns. The analysis of the editorials focused on themes of science, evidence and risk/benefit as dominant issues in new risk debates (see Beck, 1992 [1986]) and they were analysed at the four critical junctures identified above.

The Critical Interventions of Prince Charles and Professor Pusztai

Between the launch of Europe's first GM food in 1996 and June 1998, newspaper engagement with the issues was sporadic, fragmentary and ambivalent, with newspapers unsure whether to focus on benefits or risks of the novel food technology. This first changed in June 1998 with the publication of a letter by Prince Charles in the *Daily Telegraph*. In this he questioned the "rapid" expansion of GM food, articulated moral doubts about such an invasive technology, expressed lay scepticism about government reassurances about safety and raised the prospect of parallels with BSE/CJD. The mid-market *Express* and *Daily Mail* hailed this unprecedented intervention by the heir to throne as morally "right" in that the Prince was voicing the lay doubts and "fears" of the "vast majority" of ordinary people about "undue haste" in expanding GM food and crops (*Daily Mail*, 1998).

Two months later, a television documentary included preliminary research by Professor Pusztai claiming links between GM potatoes and "stunted" growth and reduced immunity in laboratory mice that were fed this produce. All four newspapers hailed Pusztai as an "independent" scientist, an expert-consumer whose own research caused him to question government certainty claims about safe food, whose moral doubts led him to question the "testing" of GM food on "human guinea pigs" and whose personal preference to avoid eating GM food was blocked by inadequate labelling policy (see *Daily Mail,* 1998; *Express,* 1998c).

The interventions of Prince Charles and Pusztai marked significant shifts in the newspapers from ambivalence to hostility. What emerged was a very powerful discourse of moral doubts, scientific uncertainty and consumer disempowerment. While Pusztai's "revelations" were seen as challenging the certainty claims of government about safe food, they did not replace these with certainty claims of harm (*Daily Mail,* 1998; *Express,* 1998c). The findings were inconclusive in that they did not "prove" that there would be the same effects in humans (*Daily Mail,* 1998). Thus, scientific uncertainty was constructed as the absence of incontrovertible evidence of safety *or* of harm, of benefit or risk—that is, a state of suspended certainty and unknowing (see *Daily Mail,* 1998; *Express,* 1998c).

The myth of objective, authoritative and unitary science collapsed in the face of conflict between scientists about what constituted credible and robust scientific methods for evaluating risk. These otherwise abstract debates were concretized in familiar foodstuffs such as bread and pizza, in parallels with government mismanagement of the contested science of BSE/CJD, in other food scares in the "history of manufactured food" such as pesticides, steroids, etc., and in cultural mythologies of "Frankenstein foods" and human "guinea pigs" (*Daily Mail,* 1998; *Express,* 1998b; *Mirror,* 1999). Thus very emotive discourses emerged out of the vacuum of suspended scientific certainty and the collapse of claims of objectivity and rationality. The deep anxiety generated by this was given added traction by a moral disquiet in "tampering" with life (*Daily Mail,* 1998) and claims about the failure of government policy to ensure proper segregation of GM and non-GM, newspapers claimed, meant a contamination of the natural, the inability of retailers to label properly and the denial of consumer choice (*Daily Mail,* 1998; *Express,* 1998b; *Mirror,* 1999).

The subsequent treatment of Pusztai was seen as vindication of his "whistle-blowing" argument. In the months after his intervention he was seen as the "independent scientist" of international standing (*Express,* 1998d), unaffected by political agendas that would subvert science, possessing the courage to raise "serious doubts" about GM food (Harrison, 1998) and "expose" food risks despite considerable professional and personal cost (*Express,* 1998d). Pusztai's employers at the Rowett Institute suspended him, claiming he had "drastically" misled the public by releasing premature findings based on a flawed methodology, banned him from speaking and then sought to discredit his work (Ingham, 1998). This was seen by the newspapers and other scientists as an attempt to "suppress" research that questioned the agenda of government and biotechnology companies.

The Launch of the Campaigns

The campaigns were a deliberate and self-conscious shift on the part of editors from classic liberal assumptions about "impartial reporting" to participatory arguments about the legitimacy of seeking to mobilize the public and influence policy in conditions of acute uncertainty, overwhelming public interest and the undemocratic tendencies of the government.

The *Express* had been the first to launch a campaign in July 1998 soon after the intervention of Prince Charles. The other three newspapers followed suit, but only after the government had rejected calls from its statutory advisor English Nature for a moratorium on the commercial cultivation of GM crops (Meacher, 1998). The campaigns were based on a tripartite agenda. First, they comprised a revelatory agenda intended to "bring out the facts" about the "hidden dangers" in food (*Express,* 1998a) and to "alert the public" to the dangers of "Frankenstein food" (*Daily Mail,* 1999). Second, an educative agenda set out to address gaps in consumer awareness about the lack of choice over whether or not to eat GM food (*Daily Mail,* 1999; *Express,* 1998a). Third, an advocacy agenda called for a change of policy in the form of comprehensive labelling and a moratorium on the commercial cultivation of GM crops (*Daily Mail,* 1999; *Express,* 1998a; *Independent on Sunday,* 1999; *Mirror,* 1999).

These three agendas combined in mutually reinforcing ways. For instance, the first two drew on assumptions that information facilitates participation, agency and power, and so holds the potential to mobilize consumers to resist policies that they disagreed with. This potential was that much greater given the readership and ideological spread to which these newspapers could appeal. This in turn could add to the credibility of the newspapers' advocacy agenda with its dual objectives of labelling and a moratorium. Thus the campaign agenda clearly articulated participatory endeavours to mobilize readers and influence policies.

The newspapers justified this departure from prevalent assumptions about newspaper behaviour in criticisms of government mishandling of the crisis and "lack of leadership" (*Express,* 1999) and the prospect of repeating the mistakes of BSE/CJD with GM food. The determination to press ahead with policy was not seen as strength but as "surreal" (*Daily Mail,* 1999) and "incomprehensible" (*Mirror,* 1999). The "explosive" rate of development was juxtaposed against the frightening spectre of suspended certainty, where there was evidence of neither safety nor harm; what "relatively little research" that had been done had "disclosed widespread possible risks" while simultaneously failing to "disclose the exact nature of risk" (*Independent on Sunday,* 1999). In this confusion, it was seen as irresponsible, irrational and illogical for government to press ahead. The time taken to gather the necessary "evidence", it was argued, could mean it would be "too late" to reverse any "malign effects" (*Express,* 1999). The papers claimed that the "real" policy agenda was not public interest but, as with BSE/CJD, a "supine" support for commercial interests and "arrogance" (*Daily Mail,* 1999). The treatment of consumers as "guinea pigs without their knowledge" (Ingham, 1998) and the "foisting" of such food and crops on the public was presented as immoral and irresponsible (*Daily Mail,* 1999). Instead of being responsive to the public, ministers were seen as "blithely indifferent" to the genuine and legitimate concerns of "millions of British consumers" (*Daily Mail,* 1999). In pushing ahead with a policy despite the "lack of consumer enthusiasm" (*Independent on Sunday,* 1999) and perceived preference of readers—"don't mess with nature" and "we want . . . GM food cleared from the shelves immediately"—it was seen as acting without a democratic mandate.

Blair Counter-attacks

Within days of the launch of the fourth newspaper campaign, Tony Blair launched a rare counter-attack in an article in the *Daily Telegraph* (Blair, 1999). He accused "parts of the media" of "scaremongering" and claimed that the campaigns signified a sudden and frenzied collective flight from rationality (a "stampede") and transformation of fact into fiction ("great GM food saga"). He counter-argued that the "extraordinary campaign of distortion" was misrepresenting or obscuring the facts, preventing the public from accessing them and so fuelling public uncertainty. The inference was that irrational, fictionalized, biased and obfuscating campaigns signified a departure from the normative notions of rational, factually accurate and impartial journalism that contributes to reasonable debate and an informed citizenry. The prime minister also dismissed claims that government was putting the public at risk, claiming instead that robust, "elaborate" and technocratic processes ensured no GM products were released for sale or cultivation unless they had been tested and verified by "independent" scientists.

Conclusion

The campaigns continued for a year and did not subside until after a *de facto* retailer boycott had emerged, government had negotiated a voluntary moratorium with the industry on the cultivation of certain types of crops and Blair had apologized, conceding that the science underpinning GM food was uncertain with no clear evidence of harm or safety, risk or benefit. That is, until the government had clearly demonstrated its responsiveness to public concerns in a way that satisfied the media. The question, then, is what we can conclude about the differences between classic liberal and participatory approaches to the media from this case study.

Classic liberalism, emerging as it did in a context of political revolution, is concerned with the role of media in creating political stability and what professional attributes are required to ensure that a free press does not undermine this. From this emerged three roles for the media: as a watchdog on political abuse when the corruption of politicians threatened to incite public anger; as provider of accurate facts so that citizens could make informed political choices at elections; and as a forum for critical-rational debate. Ideas concerning what professional skills were needed to deliver this were rooted in the enthusiasm at the time for scientific method and rationality as an antidote to undue, volatile emotions that threatened stability. These values informed the emergence of a professional emphasis on objectivity, impartiality and factual accuracy. Thus there was an assumption that the ability of journalists to "report" from a distance, dispassionately and independently, was fundamental to their ability to hold governments to account and be neutral mediators of divergent views within discursive forums of critical-rational debate.

This article has argued that these values and assumptions have been fundamentally challenged and exposed in the new risk domains that provide much of the subject matter for environmental journalism. New risk disputes are able to challenge these because knowledge is contested, scientific methods are disputed and acrimonious disagreements emerge on how to proceed in conditions of uncertainty. The ability of the journalists to engage with these issues is further complicated by, on the one hand, the abstract, hypothetical and invisible nature of the risks; and, on the other hand, the "politically catastrophic" (Beck 1992 [1986]) consequences of these and the deeply moralistic debates about the acceptability or otherwise of the risks. In the case of GM food, the bridge was constructed through critical interventions by credible figures (Prince Charles and Árpád Pusztai), in food as the embodiment of risk, in the parallels with other food scares such as BSE/CJD and in the familiar metaphoric associations of "guinea pigs" and "Frankenstein foods". The tension between the objective and polemical imperatives within newspapers' constructions of their own identity became exposed. In the case of GM food, the newspapers veered towards the polemical—a shift that classic liberalism is unable to explore adequately without a prior dismissal of it as an aberration, a deviation from professional standards of objectivity, impartiality and independence.

A participatory approach to the media as outlined here offers one normative alternative to making sense of this. Rather than being concerned with political stability, such approaches are concerned with the revival of moribund democracies and the revitalization of politics. Rather than dispassionate "observers", the public interest imperative means that the media can legitimately become active participants seeking to influence policy change in their own right and to mobilize wider political participation. Rather than a pre-occupation with great affairs of the state such as political corruption, their primary role is to expose the impact of policies on everyday activities and, contrary to the claims of political elites, highlight the disempowerment of ordinary citizens to make choices over their own lives. Rather than holding governments to account, participatory media hold them to be responsive—not in a knee-jerk way to those who shout the loudest, but in a way that includes in its consideration those views that oppose its own. Within such a framework, information is crucial—not as objective facts, but as contested knowledge and inconvenient details that those in power, both government and corporations, would rather pretend were certain or remained secret. Within the spaces of participatory media, emotion and reason become interwoven in debate rather than a duality; fear, anger, etc. can be seen as legitimate expressions of political engagement rather than dismissed for tainting the critical-rational space of debate.

The framework outlined here raises substantive questions for exploration in subsequent articles; for instance, on how to address the potential abuses of a participatory media, what happens when the forms of revitalization are so destabilizing as to risk the future of the polity, what inadequacies a critique of these approaches might highlight, and whether other approaches in modern political and democratic theory might offer more useful alternatives. The intention here was not to address these but to offer a starting point for further discussion based on the argument that environmental journalism with its new risk-type subject matter has exposed the limitations of some of the core premises in media studies, a field still overly reliant on the outdated assumptions of classic liberalism. Thus, the central contention is that it is not the redrawing or blurring of the boundaries of journalism's established demarcations that are needed but a wholesale relocation of the premises underpinning notions of the ideal role of the media and professional journalism. The issue, then, is where might the normative dimensions of media engagement with the major socio-political and economic issues of late modernity be more usefully relocated.

NOTE

1. Studies of American cases have attempted to explore the continuing tensions between environmental journalism, objectivity and campaigns (see Neuzil, 2008; Neuzil and Kovarik, 1996). However, they are of limited use here for two reasons. First, they do not specifically address the epistemological conundrums posed by the concept of objectivity within new risk debates to the subject matter of journalism. Second, their notion of campaign is assumed to be self-evident so poorly defined and analytically vague. The consequence is that distinctions are made between mainstream and specialist media campaigns but not between ones where the reader infers the publication is campaigning and one where the editors self-label this.

REFERENCES

AUSTIN, ALISON and LO, HELEN (1999) "Evidence to the House of Commons Science and Technology Committee", in: *Scientific Advisory System: genetically modified food*. London: Sainsbury's.

BECK, ULRICH (1992 [1986]) *Risk Society: towards a new modernity*, M. Ritter (Trans.), Los Angeles: Sage.

BECK, ULRICH (1999) *World Risk Society*, Cambridge: Polity.

BECK, ULRICH (2003) "Risk Society", in: M. Slattery (Ed.), *Key Ideas in Sociology*, New York: Nelson Thornes.

BIRKS, JEN (2009) "Newspaper Campaigns, Publics and Politics", PhD thesis, Glasgow University.

BLAIR, TONY (1999) "GM foods: we stand firm", *Daily Telegraph*, 20 February, p. 12.

BLAIR, TONY (2007) "Blair's Speech on the Media", *The Times*, 12 June, http://www.timesonline.co.uk/tol/news/politics/article1922074.ece?token=null&offset=0&page=1, accessed 30 November 2010.

BLUMLER, J.G. and GUREVITCH, M. (1995) *The Crisis of Public Communication*, Abingdon: Routledge.

CONBOY, MARTIN (2006) *Tabloid Britain: construction of community through language*, London: Routledge.

COTTLE, SIMON (1998) "Ulrich Beck, 'Risk Society' and the Media: a catastrophic view?", *European Journal of Communication* 13(1), pp. 5–32.

CUNNINGHAM, BRENT (2003) "Re-thinking Objectivity", *Columbia Journalism Review* July/August, http://www.alternet.org/media/16348, accessed 27 November 2010.

DAILY MAIL (1998) "Dangerous Food", 11 August, p. 10.

DAILY MAIL (1999) "An Issue of Concern to Every Reader", 6 February, p. 12.

DRYZEK, JOHN, HONIG, BONNIE and PHILLIPS, ANNE (Eds) (2006) *The Oxford Handbook of Political Theory*, Oxford: Oxford University Press.

EILDERS, CHRISTIANE (2002) "Conflict and Consonance in Media Opinion: political positions of five German quality newspapers", *European Journal of Communication* 17(1), pp. 25–63.

EXPRESS (1998) "Food Safety's Hard to Swallow", 9 June, p. 10.

EXPRESS (1998b) "The Hidden Dangers of 'Frankenstein Food'", 29 June, p. 10.

EXPRESS (1998c) "End Doubts on Food Safety", 10 August, p. 10.

EXPRESS (1998d) "Was Professor Right?", 14 August, p. 25.

EXPRESS (1999) "We Must Give a Lead Over GM Foods", 13 February, p. 10.

FIRMSTONE, JULIE (2008) "The Editorial Production Process and Editorial Values as Influences on the Opinions of the British Press Towards Europe", *Journalism Practice* 2(2), pp. 212–29.

GIDDENS, ANTHONY (1998) "Risk Society: the context of British politics", in: Jane Franklin (Ed.), *The Politics of Risk Society*, Cambridge: Polity.

GREENSLADE, ROY (2004) *Press Gang: how newspapers make profits from propaganda*, Basingstoke: Pan Macmillan.

HABERMAS, JURGEN (1989 [1962]) *The Structural Transformation of the Public Sphere*, T. Burger (Trans.), Cambridge: Polity.

HABERMAS, JURGEN (1992) "Further Reflections on the Public Sphere", in: Craig Calhoun (Ed.), *Habermas and the Public Sphere*, Cambridge, MA and London: MIT Press.

HARGREAVES, IAN (2003) *Journalism—truth or dare?*, Oxford: Oxford University Press.

HARGREAVES, IAN and LEWIS, JUSTIN (2003) *Science and the Media: towards a better map*, London: ESRC.

HARRISON, TOM (1998) "Genetic Crop Scare Scientist Suspended", *Mirror*, 13 August, p. 11.

HELD, DAVID (2006) *Models of Democracy*, 3rd edn, Stanford, CA: Stanford University Press.

HENRY, FRANCES and TATOR, CAROL (2002) *Discourses of Domination: racial bias in the Canadian English-language press*, Toronto: University of Toronto Press.

HOWARTH, ANITA (2010) "Contested Processes, Contested Influence: a case study of genetically modified food in Britain", in: Katrin Voltmer and Sigrid Koch-Baumgarten (Eds), *Public Policy and the Mass Media: the interplay of mass communication and political decision making*, London: Routledge.

HOWARTH, ANITA (forthcoming) "Newspaper Campaigns as Media Interventions", in: Kevin Howley (Ed.), *Media Interventions*, New York: Peter Lang.

INDEPENDENT ON SUNDAY (1999) "Act Now on GM", 7 February, p. 24.

INGHAM, JOHN (1998) "Professor Who Exposed Food Risk Is Sacked", *The Express*, 13 August, p. 2.

IZADI, FOAD and SAGHAYE-BIRIA, HAKIMEH (1997) "A Discourse Analysis of Elite American Editorials", *Journal of Communication Enquiry* 31(2), pp. 140–65.

KAMINSKY, JOHN P. (2006) *James Madison: champion of liberty and justice*, Madison, WI: Parallel Press.

KITZINGER, JENNY and REILLY, JACQUI (1997) "The Rise and Fall of Risk Reporting: media coverage of human genetics research, 'false memory syndrome' and 'mad cow disease'", *European Journal of Communication* 12(3), pp. 319–50.

LOCKE, SIMON (2001) "Sociology and the Public Understanding of Science: from rationalization to rhetoric", *British Journal of Sociology* 52(1), pp. 427–91.

MCQUAIL, DENIS (2000) *McQuail's Mass Communication Theory*, London, New Delhi and Thousand Oaks: Sage.

MEACHER, M. and ROOKER, J. (1998) Oral Evidence to House of Lords European Communities Committee, *EC Regulation on Genetic Modification in Agriculture*, London: TSO.

MINDICH, DAVID (2000) *Just the Facts: how "objectivity" came to define American journalism*, New York: New York University Press.

MIRROR (1999) "Label Frankenstein Food: Mirror's 'label it all' plea", 15 February, p. 2.

NEUZIL, MARK (2008) *The Environment and the Press: from adventure writing to advocacy*, Evanston, IL: Northwestern University Press.

NEUZIL, MARK and KOVARIK, WILLIAM (1996) *Mass Media & Environmental Conflict: America's green crusades*, Thousand Oaks, CA: Sage.

PATEMAN, CAROLE (1970) *Participation and Democratic Theory*, Cambridge: Cambridge University Press.

PUSZTAI, ARPAD (2002) "GM Food Safety: scientific and institutional issues", *Science as Culture* 11(1), pp. 69–92.

RICHARDS, BARRY (2004) "The emotional deficit in political communication", *Political Communication* 21(3), pp. 339–56.

RICHARDSON, JOHN D. and LANCENDORFER, KAREN M. (2004) "Framing Affirmative Action: the influence of race on newspaper editorial responses to the University of Michigan cases", *Harvard International Journal of Press/Politics* 9, pp. 74–94.

SCAMMELL, M. and SEMETKO, H. (Eds) (2000) *Media, Journalism and Democracy: A Reader*, Aldershot: Ashgate.

SCHUDSON, MICHAEL (1978) *Discovering the news: a social history of American newspapers*, New York: Basic Books.

SCHUDSON, MICHAEL (2001) "The objectivity norm in journalism", *Journalism* 2(2), pp. 149–70.

TEORELL, JAN (2006) "Political Participation and Three Theories of Democracy: a research inventory and agenda", *European Journal of Political Research* 45(5), pp. 787–92.

THOMAS, JAMES (2005) *Popular Newspapers, the Labour Party, and British Politics*, London: Routledge.

VAN DIJK, TEUN A. (1996) "Power and the News Media", *Political Communication in Action*, http://www.discourses.org/OldArticles/Power%20and%20the%20news%20media.pdf, accessed 16 February 2010.

YEARLEY, STEPHEN (1991) *The Green Case: a sociology of environmental issues, arguments and politics*, London: Unwin Hyman.

ENVIRONMENTAL JOURNALISM IN BANGLADESH

Active social agency

Jahnnabi Das

Environmental journalism challenges conventional notions in journalism, such as journalists' status as remote and neutral observers. Many environmental journalists even argue for a new position as active participants in various debates about meaning making. This position is explored in this study using Vincent Mosco's notion of "active social agency" for communication workers. By using such active agency, journalists engage in negotiations with other institutional agents in society over the meaning conveyed in news reporting about important environmental issues. This paper examines two weeks' news coverage of a media campaign in Bangladesh in June 2009; the campaign aimed to save ailing rivers around the city of Dhaka. By studying empirically which topics achieve news attention and which sources attain visibility or suffer exclusion from news content, this paper demonstrates that approaches to environmental reporting do embrace some elements of active participation in debates compared to more conventional notions of journalistic neutrality and distance. The paper also argues that the nature of environmental news reporting in Bangladesh is mainly "episodic", which is consistent with other countries, while elements of long-term "thematic" news coverage of such issues is only slightly evident.

Introduction

Bangladesh is aptly called the "land of rivers" with about 700 small and large rivers criss-crossing this low-lying South Asian country (Banglapedia, 2007). Of the hundreds of rivers, the Buriganga or "Old Ganges" has been the lifeline of the capital Dhaka. This river is now seriously polluted with high industrial and other contamination, which has severely threatened the city's public health, eco-system and socio-economic circumstances (Anam, 2009). In addition, various encroaching structures block the natural flow of these rivers. These are complex problems, some of the consequences of rapid urbanisation, lack of adequate resources in river management and blatant disregard for environmental regulations; they have been major concerns for decades (Khan, 2009). The news media in Bangladesh accordingly pay considerable and regular attention to the problems of these important rivers.

In June 2009, two media organisations—the *Daily Star*, a prominent English-language national newspaper, and Channel-i, a Bangla-language satellite television station—jointly launched a campaign titled "To the Prime Minister: Save Rivers, Save Dhaka" to highlight the severity of the problems to the newly elected government and to call for some concrete policy initiatives. The Prime Minister Sheikh Hasina and her government ministers responded by offering rhetoric, urging the need for creating greater awareness about river health and the importance of co-operating with authorities to save the rivers.

The "Save Rivers" campaign, which was conducted across two weeks from 1 to 15 June, attracted considerable attention not only from public and civil society bodies, but from other national news media. The nature of this news coverage was broadly episodic (Iyengar, 1991), which meant news stories reported either concrete instances of particular issues or were prompted by immediate events. Iyengar defines episodic news as "stories that depict issues predominantly as concrete instances or events" (1991, p. 18). In this study, this definition of "episodic" news is expanded by incorporating stories which are also engendered or triggered by immediate events or developments. Most news stories found in this particular case were somehow or other reactions to this "Save Rivers" media campaign.

This is generally consistent with the pattern of environmental reporting. Journalists often report environmental topics as "episodic" reactions (Waisbord and Peruzzotti, 2009) to something unexpected, such as natural disasters or environmental accidents, which in turn trigger additional news stories, for example, political events, statements, or rallies and demonstrations. In the case of the above media campaign, the episodic nature of reporting was evident. However, some news reporting went beyond this pattern and offered "thematic" coverage of underlying long-term issues concerning environmental problems. Iyengar defines thematic news as "stories that depict issues more generally either in terms of collective outcomes, public policy debates, or historic trends" (1991, p. 18).

In line with a previous investigation of environmental reporting in Bangladesh (Das et al., 2009), this study highlights the complex characteristics of the news-reporting pattern in an age of environmental extremities, and examines the significance of such complexities for the practice of journalism. Specifically, it studies how river pollution as a significant environmental issue in Bangladesh is depicted in the press.

Background: River Issues and Journalism

Dhaka is a global mega-city with a population of 12 million, but its infrastructures are crippled by a relentless influx of internal migrants. The city is also the heartland for all economic and socio-cultural activities of the 142.3 million people of the country (Bangladesh Bureau of Statistics, 2011). The national and international economic prospects of Bangladesh hinge on proper functioning of this city. Its environmental condition is thus of utmost national importance; so is the ecological health of the four rivers surrounding the city—the Buriganga, the Shitalakhaya, the Balu and the Turag. These rivers provide drinking water besides facilitating vital irrigation, transportation and local-level fishing.

However, these rivers are increasingly becoming "biologically dead" because of the dumping of untreated industry and human wastes, the encroachment of river banks and the construction of structures obstructing the rivers' natural flows (Khan and Roy, 2011). The problems of pollution and river encroachment are prevalent in all parts of Bangladesh, but are particularly severe in the rivers around Dhaka. Consequently, civil society bodies and experts have been calling to save Dhaka by saving its rivers. In these circumstances, the protection of rivers deservedly receives extensive news coverage.

Journalism in Bangladesh is vibrant with newspapers, broadcasting and new media operations in both private and public sectors. Contrary to the recent decline of print media in the Western world, newspapers in Bangladesh are still thriving and playing a strong role in politics and society. According to the official statistics, there are 467 newspapers in Bangladesh including 292 dailies; but only dozens of these are substantial and significant publications. Circulation figures are increasing, which is consistent with the overall trend in other Asian countries (Department of Films and Publications, 2010; Merrill, 1995; World Newspaper Association, 2010). This indicates that newspapers remain an important element of journalism in Bangladesh, and deserve continued scholarly attention.

The role of journalism in this country has been complex, and the two polarised positions of "neutral" and "participant" roles (see Hanitzsch, 2005) for journalists seem to be inadequate to describe it. Historically, newspapers in Bangladesh have been intricately connected to various prolonged political movements, such as, gaining independence from the colonial British and Pakistani rules in 1947 and 1971, respectively, and establishing democracy by overthrowing a military dictatorship in 1990. Many journalists are strongly committed to "social change" and ready to move beyond the professional mindset of the distant observer and neutral reporter to intervene in any situation that requires action. This, however, does not weaken their commitment to the liberal ideals of "objective" journalism by keeping themselves above the immediate political fray and upholding professional standards of accuracy, fairness and balance (Anam, 2007; Rahman 2005; Ramaprasad and Rahman 2006).

Environmental Journalism: A Brief Review of Relevant Literature

Scholarly investigations into environmental journalism (Brossard et al. 2004; Hansen, 1993; Hutchins and Lester, 2006; Neuzil, 2008; Shanahan and McComas, 1997; Spellerberg et al., 2006) are significant in number, but mainly Western-centric, which is quite consistent with research in other aspects of media and journalism (Josephi, 2005). Some recent studies (Eide et al., 2010), however, are addressing this gap by focusing on environmental journalism in non-Western contexts, for instance, from South-East Asia (Nash and Bacon, 2006) or Latin America (Guedes, 2000; Waisbord and Peruzzotti, 2009). Findings from these studies indicate that environmental reporting in these contexts adopts practices which are also familiar in the West, such as resorting to "episodic" coverage and "staying in the comfort zone of covering the political game" surrounding environmental problems.

Environmental journalism has always been considered a specialised area, which requires a level of expertise and deep understanding of relevant issues (Schwartz, 2006; Sharma, 2010). This requirement of expertise generates intense debates in both scholarly literature and professional circles about the preferred way of practising such journalism, which often requires journalists to raise public awareness about environmental issues and make subjective judgements on contentious questions. Perhaps that is why a US-based professional body—the Society for Environmental Journalists—has eliminated "objectivity" from its code of ethics, asserting that the organisation should not police its members against reflecting their personal convictions in news reports (Palen, 1999). This obviously makes environmental journalism complicated, because the conventional view implies that journalists as independent reporters need to guard their facts from values and other vested interests (Nelkin, 1995; Palen, 1999; Schudson, 1978).

This debate also finds its way into various professional deliberations about journalism in Bangladesh. In a recent discussion, some journalists argued that environmental reporting should come under one broad umbrella of "journalism" which needs to be balanced and objective to protect the independence of this practice (*Prothom Alo*, 2010). However, others (Anam, 2007; Bavadam, 2010) consider the scale of environmental and associated problems in South Asia, including Bangladesh, to be so considerable that they require journalists to be driven by moral imperatives rather than professional norms on this issue; they must take pro-active roles to uplift the existing underdog status of environmental reporting and intervene in situations. In short, environmental journalism necessarily involves an active agency.

Theory: Journalistic Agency

Many critics perceive environmental reporting as advocacy journalism (Neuzil, 2008; Wyss, 2008) in which journalists actively participate, interpret, and speak on behalf of those who have been denied access to media representation and power. The broad purpose of such reporting is to address the "power imbalance" in society (Janowitz, 1975; Waisbord, 2009). Such active participation of journalists in the news-making process sits oddly with the mainstream watchdog model of journalism guided by professional ideals of objective, fair, balanced and detached reporting (Schudson and Anderson, 2009; Waisbord, 2009). However, environmental reporting, as part of the field of journalism, is also perceived as inextricably intertwined with truth, fairness, balance and neutrality (Lichtenberg, 1999, p. 225). These attributes make journalism distinct from other similar practices in society, such as politics or literature, and enable journalists to claim they discover and represent the social world in their own particular ways (Neveu, 2007).

This differentiation of journalism from other institutions in society (Hallin, 2005) gives journalists the power to negotiate with the representatives of other institutions in news production processes. Journalists exercise this power by verifying facts from different sources and presenting such facts as authentic in news content (Ericson et al., 1989; Tuchman, 1978). This is why the relationship between journalists and sources is highly significant in news production (Franklin et al., 2010). This relationship is sustained through the requirement of objectivity and balance by maintaining a "web of facticity" (Tuchman, 1978), which requires journalists to validate facts from sources. The adherence to norms of "facticity" allows news media to endorse truth claims and also "suggest their preferred meanings" of any given reality (Nash and Bacon, 2006, p. 109; also Ericson et al., 1989). Such processes of adherence and validation are also crucial in the reporting of environmental issues, and this may enable reporters to advance the cause of the environment. At the same time, this validation and adherence raise questions about who controls the preferred meaning of news.

On this question, Ericson et al. (1989, pp. 1–7, 13–16) contend it is not a straightforward or unilateral control of source organisations over media organisations in reproducing their worldview in the public domain. Instead, sources and journalists are constantly engaged in a "struggle over the control of accounts" in which journalists as the authoritative apparatus have power over their sources or "authorised knowers" to structure events and select sources for their readers. Accordingly, the journalist has the power to articulate the final or last word.

But, as indicated above, this power of the final word is not guaranteed, because in the production of news,

> What is at stake . . . is the meaning attributed to events, processes or state of affairs . . . There are perpetual conflicts over preferred signs and appropriate meanings . . . Negotiation of control over signs and meanings occurs at every stage: among members and units within a particular source organization; among reporters and editors within a particular news organization; between members of different news organisations; and between news organizations and source organizations. (Ericson et al., 1989, p. 377)

The multi-pronged contestation indicates that although journalists have the power of the final word, it is not always given, rather more contingent. In the investigation of environmental news production, then, an important matter is the continuous contestation over meaning.

This is where the idea of "active social agency" (Ballesteros et al., 2010, p. 4; Mosco, 2009) becomes relevant for journalism in general and environmental reporting in particular. By this term, Mosco meant the active participation of communication workers, including journalists, in the process of creating value in contemporary society. He asserted that in much of the research about labour, media and technology, the main concern had been the exploitation and degradation of labour at the hands of powerful corporations, which was very important but not the whole story about creative labour. Mosco further asserted that communication labour also actively participated in the creation of value. Part of such participation involved resistance to corporate dominance in various ways. In the case of journalism, this means bringing up and highlighting issues and matters of content that go against the interests of large corporate entities. Environmental questions are one such instance. Despite the continued struggle between sources and news organisations over the meaning and definition of legitimate news, journalists and editors increasingly established environmental issues as important and legitimate news topics.

This study considers the source analyses of Ericson et al. (1989) and Nash and Bacon (2006) as instructive in advancing understanding of the complex web of interactions between news sources and journalists.

As most newspapers in Bangladesh adopt a default position of being an agent of "social change" for public good (Anam, 2007; Rahman, 2005), the main concern of the current study is to ascertain which river-related issues receive most news attention, and which types of sources authenticate news reports and legitimise journalists' position as agents of change in society.

In order to examine the authentication and legitimation of the default position of social change, this study examines "what is known", i.e. what themes are selected from the range of environmental issues; and "how it is known" (Tuchman, 1978), i.e. how sources are included and excluded, as well as contextualised, in the coverage of river issues in selected newspapers.

Methodology: Content Analysis

A content analysis of news items on river issues from three newspapers was conducted during the "Save Rivers" campaign (1–15 June 2009) (Figure 1). The

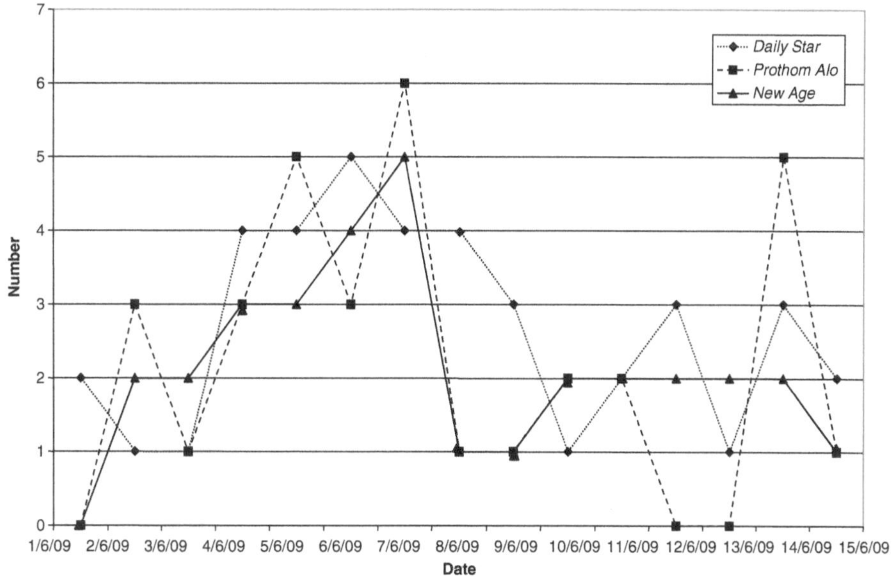

FIGURE 1
Coverage (number of items) of river pollution in three newspapers

publications were: the Bangla-language daily the *Prothom Alo* (circulation 406,525), and the English-language papers the *Daily Star* (campaign organiser, circulation 40,262) and the New Age (circulation 15,430) (Department of Films and Publications, 2010).

The selection of publications offers the opportunity to examine diverse perspectives in news coverage: the *Prothom Alo* is the highest-circulating and most prominent newspaper in the national language; the *New Age* is a very recent publication representing new talents and perspectives in the country's journalism, while the *Daily Star* is a prestige daily with limited circulation but considerable clout over the policy circle because of its elite readership.

The time of the data collection allows examination of the river-related issues reported during the campaign, but also enables a comparison with the previous study (Das et al., 2009), which also collected news items in the month of June but two years previously (2007). The timing of the study is significant in another way: in Bangladesh, the political scenario had changed in early 2009 as a newly elected government assumed office following two years of turbulence under a non-elected interim administration.

The content analysis was conducted in two stages: firstly, a broad mapping of items in the selected newspapers was conducted to identify the issues at stake about the river. Then, a more detailed numerical and qualitative examination was conducted to obtain deeper insights into the identified issues. Copies of items including news, editorial and opinion pieces had been obtained through online archives of the newspapers. Electronic database key term search was not available for the Bangla-language *Prothom Alo* newspaper, so a manual scanning of online pages was undertaken by the author who possesses the necessary language skills. In addition to the content analysis, data from telephone interviews with some environmental journalists in Bangladesh have been used to add further depth to the discussion.

Findings

The findings are organised and presented in two sections. The first describes general themes that emerged in the media coverage of river pollution issues, and in the second, characteristics of news sources, verification of statements made by sources and the attribution of responsibility for environmental wrong-doings are detailed.

During the two-week span of the study, a broad mapping exercise identified an aggregate 185 river-related items across the three sample newspapers; 105 (85 per cent) of these stories were about river pollution and related matters, 42 (23 per cent) about ecological issues, 13 (7 per cent) about natural disaster, seven (3 per cent) about climate change and 18 (9 per cent) were about "others", a residual category which included the repair of river embankments and problems of fishing in these rivers (Table 1). Of these, river pollution and related items (105) were directly linked to the research question and subjected to further in-depth qualitative and quantitative analyses (Table 2).

Themes

The main themes that emerged in the analysis of newspaper contents included the call for protection of the rivers, water contamination of the river, land filling and encroachment of the river, and the eviction drive designed to free the river of illegal occupants. These themes are elaborated below.

Protection. During the two weeks of the study, a major concern for these newspapers had been to report on calls for protection of the rivers from environmental degradation. These calls were made at different discussion seminars and political rallies, mainly by two groups of people: first, the non-government organisations urging the government to take immediate action to address various river problems, and second, government ministers urging "everyone" to create more awareness about river health and promising to provide improved administrative and legal support to resuscitate the dying rivers.

The significance of reporting these "calls" is that such news items "mandate" (Tuchman, 1978, p. 134) river problems as a valid news hook, and an issue of broad social concern. The editorial decision to publish these "calls" as news reports also indicate a change in the reporting approach: here, journalists are not maintaining conventional journalistic neutrality; instead they are adopting a campaign mode for environmental causes. This explains why these calls, which are otherwise insignificant as news events, occupy important positions in the pages of newspapers.

Water contamination. During the media campaign, these newspapers tried to move beyond the "episodic" coverage pattern to adopt a "thematic" reporting style. This was evident in some news reports which dealt directly with problems instead of just reporting

TABLE 1
Coverage (number of items) of river issues in the three newspapers

Newspaper	Pollution	Ecology	Natural disaster	Climate change	Others	Total
Daily Star	40	16	2	2	3	63
Prothom Alo	33	12	8	2	14	69
New Age	32	14	3	3	1	53
Total	105	42	13	7	18	185
Percentage	56.7	22.7	7.0	3.7	9.7	

TABLE 2
Themes about river pollution in the three newspapers

Newspaper	Encroachment % (N)	Eviction % (N)	Protection % (N)	Water contamination % (N)
Daily Star	17.5 (7)	22.5 (9)	47.5 (19)	12.5 (5)
Prothom Alo	33.3 (11)	33.3 (11)	24.2 (8)	9.0 (3)
New Age	9.4 (3)	34.4 (11)	43.8 (14)	12.5 (4)

the above "calls" or other stage-managed news events initiated by various organisations. One such problem was water pollution. River water has been polluted as untreated industrial effluents and other wastes are indiscriminately and illegally dumped in these rivers. This problem is endemic across the country, as evident in a report published in the *Prothom Alo* about river pollution in the South-eastern Chittagong region:

> The health of seven lakh [0.7 million] people in Rangamati [a hill district in the country's south-east] is endangered due to indiscriminate dumping of human waste in the Kaptai lake, the only source of water for the city. (Chakma, 2009)

The *Daily Star* offered an equally important opinion piece by an expert about pollution of rivers around Dhaka, which showed the severity and extent of pollution in these waters. The article held industries, particularly the tanneries in the city outskirts, responsible for polluting the Buriganga water, and demanded concrete government actions and policies to tackle this problem. Similarly, the *New Age* (2009) published an editorial praising the government for its initiative in tackling falling ground water level by treating and decontaminating river water and making it consumable again. Contamination of both ground and surface waters has long been an environmental concern in this country.

Encroachment of the river. Another problem hampering river health is illegal occupation of the riverbanks and the construction of unauthorised structures. Influential individuals and businesses somehow occupy chunks of lands belonging to river, earth fill the encroached area, build structures and establish ownership. This hampers the natural flow of water and disrupts the river's natural ecology.

This issue was a matter of particular attention in news coverage during the two weeks of study. Of the total news items about river issues, seven in the *Daily Star*, 11 in the *Prothom Alo* and three in the *New Age* are about land encroachment. The numbers indicate that journalists intended to attach considerable editorial salience to this issue by exposing the group of "unscrupulous people" who are directly involved in earth filling and thereby the "disappearance" of the river.

The eviction drive. Eviction drives are initiatives of various government departments to evict "illegal occupants" from lands which they assert belonged to the rivers. The drives typically involve demolition of unauthorised structures. Judging by the number of stories on this topic, it seems the three newspapers also considered the eviction drive as significant (Table 2). Although the *Prothom Alo* and the *New Age* had the same number of stories about eviction, the *Prothom Alo* journalists came up with more diverse perspectives in their reports. The news items in this newspaper were different from the others in two main ways: first, the Bangla newspaper endeavoured specifically to name the so-called "land grabbers", evicted individuals or businesses (often manufacturing factories). Second, it focused on the post-eviction restoration process of the freed land. Some reports even raised questions about the efficacy of eviction drives, raising doubts about the authorities' ability to stop the perpetrators of land grabbing. One news report quoted a local source saying:

> they [the "land grabbers"] built the buildings, they built the walls, once the eviction drive is over, they will start again rebuilding on the river bank; nobody will ask to stop them, not even the police. (Dutta, 2009)

Despite this critical attitude towards the authorities' initiatives in some news items, the overall tone of the coverage was positive and in favour of the government. Broadly, the news reports studied were supportive of such initiatives with each of the three newspapers adopting the editorial position that immediate and drastic action was needed to save the rivers around Dhaka. Thus, the eviction stories in the early days of the campaign contained meticulous numerical statistics of evicted "illegal" structures reinforcing the government position that this administration was sincere about improving the river condition. The newspapers had reason to be pro-active, positive and confident about such coverage, as environmental journalist Morshed Ali Khan explained:

> Dhaka is one of the most polluted cities in the world. As news media, we need to act. We acted before, that's how we stopped [importing of] three-wheel taxis with two-stroke engine, which was responsible for heavy lead pollution in the capital. (interview, 15 June 2009)

So, the argument ran, news reporting supportive of government initiatives was reasonable and necessary. Such positive emphases on government actions can be explained and understood as newspapers' efforts to legitimise the authority's policies on river protection. The news coverage, however, did not remain positive as the eviction drive unfolded. Gradually towards the second week of the campaign, eviction-related stories started to include the post-eviction aftermath and focused on restoration of the re-possessed lands. The *Prothom Alo* story, mentioned earlier, is an example of such reporting.

Sources and Verification

While the description of different themes provides a context to the river issues in general, an analysis of news sources sheds light on the complexities of the coverage of river pollution in the three newspapers. Altogether, seven categories of sources have been identified for all direct and indirect quotes in the selected news items; they are: politicians, bureaucrats, and alleged perpetrators of environmental problems, non-government activists, experts, lay sources, and anonymous sources (Table 3).

Analysis reveals that government bureaucrats in different departments are dominant as sources in river-related news. This is consistent with the findings of a previous study of Bangladesh's environmental news (Das et al., 2009). However, a significant difference between the two is that compared to the previous research, the current study records more political sources and less non-governmental activist sources in the selected news items. A high presence of political sources is observed across the three newspapers; in the *New Age*, 20 per cent of the sources are politicians; with slightly lower figures in the *Daily Star* (17 per cent) and the *Prothom Alo* (16 per cent).

The greater visibility of political sources can be explained by the fact that these news items were published in 2009 when a newly elected government took over from a

TABLE 3
Different types of sources that appeared in the three newspapers

Newspaper	Political	Bureaucratic	Perpetrator	Activist	Expert	Lay	Anon.	Other
Daily Star								
%	17.5	53.4	3.9	8.7	3.9	2.9	5.8	3.9
N	18	55	4	9	4	3	6	4
Prothom Alo								
%	15.6	32.2	17.8	3.3	7.8	12.2	8.9	2.2
N	14	29	16	3	7	11	8	2
New Age								
%	19.5	34.1	0	21.9	7.3	0	2.4	14.6
N	8	14	0	9	3	0	1	6

non-elected interim caretaker administration, which temporarily "banned" all political activities in 2007, a period when news items for the previous study had been published. The selection of more political sources by journalists this time around, compared to more bureaucratic sources in 2007, demonstrates that journalists possess the important ability to select their sources according to what they believe to be important, significant and legitimate.

In this way, they tend to be one of the key participants in the political process of representation. However, an argument can also be made that journalists are not "actively selecting" their sources, but rather simply responding to a new political reality which emerged in the country. In a regime with an elected government, politicians are indeed important in decision making, and journalists need to pay adequate attention to them. Hence, more political sources are quoted in news, compared to bureaucratic or other types of sources than in previous times.

However, the analysis showed a negative co-relation in general between political sources and "verification" of statements: more political sources equalled less verification of statements (Table 4). Verification here means journalists critically examining any contestable statement made by sources. Whether an item is verified or not was determined by going through the item carefully and looking for signs of critical examination (such as cross-checking of stated facts with other sources; raising questions about the logical plausibility of sources' claims).

TABLE 4
Verification of statements by the sources in the three newspapers

Source	Newspaper	Verified	Non-verified
Bureaucrat	*Daily Star*	9	3
	Prothom Alo	15	2
	New Age	6	3
Politician	*Daily Star*	2	6
	Prothom Alo	3	3
	New Age	1	4
Total		36	21

This negative co-relation suggests that journalists generally reproduced the reality of river problems according to the views of politicians. However, caution is necessary in making such an observation because the news items here are mainly daily routine coverage, which usually contain less verifiable statements or have little scope for a journalist to verify a claim unless done by sources themselves.

This study further found a positive co-relation between "verification" and the use of "bureaucratic" sources, which meant more bureaucratic sources in news equalled more verification of facts by journalists. Government bureaucrats are powerful agents and significant suppliers of raw materials for news. In the case of this study, these bureaucratic sources appeared more in "eviction"- and "land grabbing"-related stories. This finding is consistent across the three newspapers.

However, there is a twist in this finding: on the river encroachment issue, these papers portrayed the bureaucratic version of reality and largely ignored cross-checking or verification of claims with the alleged perpetrators. Some bureaucrats even highlighted the difficulties of the eviction drive, such as lack of resources and legal complications; but journalists did not follow those leads and explore the statements further. Rather, in some items, they presented the bureaucratic version of the situation in an uncritical manner and provided the numerical accounts of demolished structures instead of going and talking to the alleged wrongdoers.

An important aspect of source selection in this study, especially in the two English dailies, is they used very few alleged perpetrators of land grabbing and pollution as sources. However, the Bangla daily, the *Prothom Alo*, cited a high percentage of "perpetrator" sources (17.8 per cent) while the *Daily Star* had only 3.9 per cent and the *New Age* had none. The *Prothom Alo* also had the highest percentage of "verified" news items.

As Table 4 shows, this study identified 57 verifiable news items out of total 105 items related to river pollution issues. The 48 non-verifiable items included 17 non-news items such as letters to the editor, opinion pieces and photo items with captions, and 31 items considered as general daily news about river pollution such as political rallies and meetings in which journalists themselves did not initiate the item and merely responded to the organisers' initiatives. In such items, the scope of exercising critical verification by journalists was quite limited for practical reasons including lack of sufficient time. Of the 57 verifiable items, 36 were adequately verified while in the rest of the 21 items, the journalists did not contest various claims made by the sources.

In numerical terms, journalists did not try to verify any contestable statements (and thus did not use their active agency) in the case of 52 potential items, but did so in 36 other items. The 52 potentially contestable items includes 21 non-verifiable items and 31 general daily news stories, but excludes 17 non-news items (letters to the editors, etc.). However, of the items suitable for verification (i.e., 57), the journalists did verify more items (36) than those left unverified (21). Overall, among all river-related news items, the number of verified items was less than the non-verified items. This finding is consistent with existing research about cross-checking in journalism. For example, Franklin et al. in their recent examination of public relations materials in "quality" news content in the United Kingdom found that journalists rarely attempted to contextualise news items with more than one source, and "when stories were based on specific factual claims", they discovered, "that on 70% of occasions these claims were entirely uncorroborated and in only 12% of cases were they corroborated completely" (2010, p. 208).

In the case of the river coverage in Bangladesh, journalists fared better than the UK "quality" national press, because they verified more news items (63 per cent of items, or 36 out of 57, deemed verifiable by this author; Table 4). In terms of the overall total "news items" (88), the journalists in Bangladesh verified 40 per cent, which is again greater than the UK figure of 12 per cent complete corroboration.

However, a significant qualification should be made here, because the current study examined a specific media campaign in which news organisations invested a good deal of effort to produce relatively few quality news items with high professional standards of accuracy and critical examination. By contrast, the UK study included all general news items to measure the overall ability of news organisations to maintain high-quality journalism. This represents a more reliable overall picture of reality in journalism about verification or corroboration than the current study.

Overall, the news coverage was highlighting—and critically—influential people and holding them responsible for the degradation of the health of the river. In this light, the absence of "perpetrator" sources questions the ideal of balance and fairness in news reporting, and thus impinges on journalists' critical capacity. When asked for comment about the quality of the reporting in environmental news, a senior journalist of Bangladesh, Farid Hossain, said:

> These reportings are not always up to the expected professional standard. Some of them are indeed lacking in terms of accuracy and balance. Other than general weakness found in other types of reporting, another reason for this to occur in environmental reporting is that many of the journalists involved in such reporting take it as an issue of activism and less willing to accept views from other side of the argument. They think as they are doing this sort of reporting for a good cause, overlooking some standards of reporting is justified. So the standard of reporting is not always maintained. (interview, 8 June 2009)

The use of anonymous sources was another aspect of the news items examined in this study. The *Prothom Alo* had the highest such sources in its content (9 per cent). Although anonymous sources question the credibility of news reporting, this needs to be understood in the context of endemic corruption and malpractices in Bangladesh. Many sources are unlikely to come forward and be named but still want the information to come out in public.

Journalist Morshed Ali Khan, who has been reporting environmental issues for decades and is an important organiser of the "Save Rivers" media campaign, was asked about the issue of verification and the use of anonymous sources in news reports. He asserted the use of anonymous sources was sometimes part of the strategy journalists had to adopt to get issues raised in the public space. Khan claimed although "corruption is one of the major problems here" (in Bangladesh), reporters often had to improvise because of lack of access to information: "We don't have enough information; so we do visual judgement [of the situation] rather than scientific judgement. It is so complicated."

He said sometimes even high officials in government departments instructed their subordinates to pass information about malpractices to journalists for exposure. "So it's my duty to protect these sources. I would call them humble blowers instead of whistle blowers" (Khan, interview, 15 June 2009).

By "humble blowers" he meant that the sources are unwilling or not courageous enough to be whistle blowers in the conventional sense; that is, open and named insiders. Rather, they wish to keep themselves away from the limelight, but still want to make the information public.

Attribution of Responsibility

This theme was identified to examine which individuals or organisations were held responsible, or what reasons were attributed in the newspapers, for the degrading river conditions. Five factors or categories were identified as being responsible for the river problems: the government, "perpetrators" of encroachment, legal complexities, uncertain or unclear responsibility, and "other" (Table 5). The results indicated the three newspapers almost unanimously held the "perpetrators" (i.e. influential individuals or businesses) responsible to a great extent for the pollution and encroachment of river land. The coverage also emphasised the tardiness or inaction of government departments in tackling river problems as an important reason for degradation. Further, complexities in legal procedures relating to the eviction of unauthorised structures also drew some attention. This often put the government in a difficult situation, because its efforts to recover lands are frustrated by court challenges by the "perpetrators" against eviction orders.

Frustrating legal procedures also affect journalists' approach to a problem, as explained by journalist Farid Hossain:

> We have weaknesses in our legal system; as a consequence... issues including environmental ones cannot be solved satisfactorily through the legal systems where often cases remain pending for long periods of time. So journalists think since legal remedy is not effective in preventing environmental wrong-doers, lets expose the culprits in as many ways as we can, and see if anything happens. (interview, 8 June 2009)

The attribution of responsibility turned out to be more meaningful when examined in conjunction with the presence of other sources and "verification" of statements. Although newspapers were holding the "perpetrators" responsible for river pollution and other ills, perpetrator sources were notably absent in the two English dailies. The *Prothom Alo* included some of those sources in its news content to highlight the other side of the debate.

Generally, the low visibility of "perpetrator" sources was matched by the low number of verified items. While the identification of the "perpetrators" as responsible was dominant in content, there were a significant number of items in which no individual, business or reason was held accountable for deteriorating river conditions. These items were categorised as "not clear". The presence of such items showed that the matter of identifying responsibility was not a straightforward affair; rather, in some instances,

TABLE 5
Attribution of responsibility to different organisations/individuals in the three newspapers

Newspaper	Perpetrators	Government	Legal	Not clear	Others
Daily Star	22	4	1	10	5
Prothom Alo	16	4	1	12	0
New Age	14	5	2	11	0
Total	52	13	4	33	5

determining the responsibility of environmental degradation of the rivers was difficult. Despite that, journalists' identification of responsibility in some news items indicated that they had been actively using their agency in shaping the coverage pattern on this issue.

Discussion

In covering the "Save Rivers" campaign, the two English-language newspapers preferred to put significant emphasis on the call for the protection of rivers, while the Bangla-language *Prothom Alo* preferred to highlight the issue of land grabbing and eviction. The newspapers' emphasis on the call for the protection of rivers might be explained by their preference to deal with policy issues.

The high number of political and bureaucratic sources and low "perpetrator" sources might be taken as reflecting journalists' ambivalence about explicitly identifying any party responsible for the problems. However, the actual question here was why did journalists not offer a balanced account of both the authorities and the perpetrators? One potential reason is that the pollution of the river was so severe and endemic that it did not require or deserve a balanced reporting.

The pattern of reporting showed news coverage had gradually shifted from a positive to an adversarial approach to the authorities by exposing the complexities and mixed success of the eviction drive. This demonstrated that even though the news media had broadly committed to a fairly uncritical attitude towards the government's actions, this was not a guarantee of continued favourable news coverage. When journalists found that the positive news portrayal was not effective in pursuing their cause of protecting the rivers, they shifted their focus on to the inefficacies in the eviction drive (the perceived solution to the problems).

This study also refined the previous study on environmental news reporting in Bangladesh by showing that the news media adopted different approaches to environmental issues in the case of these two studies. In both instances, the news media tried to intervene into the public policy domain, but did so in different ways. The previous study found that the journalists in Bangladesh allied with non-government organisations to challenge the government in pursuing their aim of protecting the environment. By contrast, the current study demonstrated that such an adversarial approach towards government had been transferred to some kind of co-operation: the authorities and journalists were now partners in pursuing the same goal of environmental protection.

One potential explanation of such a shift is that there was a change in the country's political environment, that the importance of politicians in decision making had increased significantly with a new government, that the interim administration had a different agenda to an elected government, and the news media were positive this time because the elected public representatives were more responsive and responsible towards the public good and environmental causes.

Conclusion

Overall, this study demonstrates that river pollution and related issues are not purely environmental questions to be dealt with administratively or technically by experts. These are complex political and social issues, and news media are an important party to the negotiation processes about these issues. This explains why journalists make conscious decisions about news topic and source selections: they include and exclude particular types of stories and sources to position themselves as "active agents" in the policy negotiation process.

In response to the research questions, it seems the main focus of attention of river news was on the call for the protection of the rivers and the eviction drive to free the encroached lands. The main types of sources were political leaders and bureaucratic officials, with a deliberate exclusion of those individuals or businesses deemed responsible for environmental ills of the rivers.

These empirical findings clearly indicate journalists' active constitution of a social problem (Ericson et al., 1987). By identifying the problem and highlighting the potential solution to it, journalists here legitimise their role as agents of "social change". As active agents (Mosco in Ballesteros et al., 2010), they adopt and change their approaches to news to align with the changing contexts indicating the fundamentally fluid nature of the practice of journalism.

ACKNOWLEDGEMENTS

I wish to thank Dr. Henrik Bødker of Aarhus University, Dr. Irene Neverla of the University of Hamburg and Akteruz Zaman of Monash University for their valuable comments.

REFERENCES

ANAM, MAHFUZ (2007) "Role of the Media on Democratic Ownership of the Development Finance System", 27 September, www.oecd.org/dataoecd/3/45/39364379.pd, accessed 11 October 2009.

ANAM, MAHFUZ (2009) "Death of Rivers Will Lead to Death of Our Cities", *Daily Star*, 1 June, http://www.thedailystar.net/newDesign/save_river/news2.php, accessed 5 November 2009.

BANGLADESH BUREAU OF STATISTICS (2011) "Statistics Bangladesh, Government of Bangladesh", Dhaka", http://www.bbs.gov.bd/Home.aspx, accessed 20 October 2011.

BANGLAPEDIA (2007) "River", http://www.banglapedia.org/httpdocs/HT/R_0207.HTM, accessed 20 October 2011.

BALLESTEROS, BEGOÑA, LUJÁN, LUIS and PEDRO, JOAN (2010) "The Political Economy of Communication: power and resistance, an interview with Vincent Mosco", *Global Media Journal* American edition, 10(17), Fall, http://lass.calumet.purdue.edu/cca/gmj/fa10/gmj-fa10-article2-ballesteros-lujan-pedro.htm, accessed 23 March 2011.

BAVADAM, LYLA (2010) "Environmental Stories Among the Most Challenging", in: Keya Acharya and Frederick Noronha (Eds), *The Green Ink: environmental journalism in India and South Asia*, New Delhi: Sage, pp. 3–11.

BROSSARD, DOMINIQUE, SHANAHAN, JAMES and MCCOMAS, KATHERINE (2004) "Are Issue-cycles Culturally Constructed? A comparison of French and American coverage of global climate change", *Mass Communication and Society* 7(3), pp. 359–77.

CHAKMA, HORIKISHORE (2009) "Public Health Threatened by Kaptai Lake Pollution", *Prothom Alo*, 5 June, http://www.prothom-alo.net/V1/archive/news_details_mcat.php?dt = 2009-06-05 &issu..., accessed 29 October 2009.

DAS, JAHNNABI, BACON, WENDY and ZAMAN, AKHTERUZ (2009) "Covering Environmental Issues and Global Warming in Delta Land: a study of three newspapers", *Pacific Journalism Review* 15(2), pp. 10–32.

DEPARTMENT OF FILMS AND PUBLICATIONS (2010) "Media List Newspaper", http://www.dfp.gov.bd/en/index.php?option = com_jdownloads&Itemid = 38&task = view.download&cid = 1, accessed 28 November 2010.

DUTTA, ARUP, (2009) "130 Illegal Structures Evicted, But No Restoration of Earth-filled Land", *Prothom Alo*, 6 June, http://www.prothom-alo.net/V1/archive/news_details_home.php?dt = 2009-06-06&iss..., accessed 29 October 2009.

EIDE, ELISABETH, KUNELIUS, RISTO and KUMPU, VILLE (2010) *Global Climate–Local Journalisms: a transnational study of how media make sense of climate summits*, Bochum: Projekt Verlag.

ERICSON, RICHARD V., BARANEK, PATRICIA M. and CHAN, JANET B. L. (1987) *Visualizing Deviance: a study of news organization*, Toronto: University of Toronto Press.

ERICSON, RICHARD V., BARANEK, PATRICIA M. and CHAN, JANET B. L. (1989) *Negotiating Control: a study of news sources*, Milton Keynes: Open University Press.

FRANKLIN, BOB, LEWIS, JUSTIN and WILLIAMS, ANDREW (2010) "Journalism, News Sources and Public Relations", in: Stuart Allan (Ed.), *The Routledge Companion to News and Journalism*, Abingdon: Routledge, pp. 202–12.

GUEDES, OLGA (2000) "Environmental Issues in the Brazillian Press", *Gazette* 62(6), pp. 537–54.

HALLIN, DANIEL C. (2005) "Field Theory, Differentiation Theory, and Comparative Media Research", in: Rodney Benson and Erik Neveu (Eds), *Bourdieu and the Journalistic Field*, Cambridge: Polity Press, pp. 224–43.

HANITZSCH, THOMAS (2005) "Journalists in Indonesia: educated but timid watchdogs", *Journalism Studies* 6(4), pp. 493–508.

HANSEN, ANDERS (1993) *The Mass Media and Environmental Issues*, Leicester: Leicester University Press.

HUTCHINS, BRETT and LESTER, LIBBY (2006) "Environmental Protest and Tap-dancing With the Media in the Information Age", *Media, Culture & Society* 28(3), pp. 433–51.

IYENGAR, SHANTO (1991) *Is Anyone Responsible? How television frames political issues*, Chicago and London: University of Chicago Press.

JANOWITZ, MORRIS (1975) "Professional Models in Journalism: the gatekeeper and the advocate", *Journalism Quarterly* 52(4), pp. 618–26.

JOSEPHI, BEATE (2005) "Journalism in the Global Age: between normative and empirical", *Gazette* 67(6), pp. 575–90.

KHAN, MORSHED A. (2009) "Buriganga Artery Blocked", *Daily Star*, 8 June, http://www.thedailystar.net/story.php?nid = 91654, accessed 25 October 2009.

KHAN, MORSHED A. and ROY, PINAKI (2011) "2 Years After: rivers still on deathbed", *Daily Star*, 2 August, special edition, pp. 1–4.

LICHTENBERG, JUDITH (1995) "In Defence of Objectivity Revisited", in: Michael Gurevitch and James Curran (Eds), *Mass Media and Society*, London: Arnold, pp. 225–42.

MERRILL, JOHN C. (Ed.) (1995) *Global Journalism: a survey of international communication*, New York: Longman.

MOSCO, VINCENT (2009) *The Political Economy of Communication*, 2nd edn, Los Angeles: Sage.

NASH, CHRIS and BACON, WENDY (2006) "Reporting Sustainability in the English-language Press of Southeast Asia", *Pacific Journalism Review* 12(2), pp. 106–35.

NELKIN, DOROTHY (1995) *Selling Science: how the press covers science and technology*, New York: Freeman.

NEUZIL, MARK (2008) *The Environment and the Press*, Evanston, IL: Northwestern University Press.

NEVEU, ERIK (2007) "Pierre Bourdieu: sociologist of media, or sociologist for media scholars?", *Journalism Studies* 8(2), pp. 335–47.

NEW AGE (2009) "Pollution of Rivers Must Stop to Make Water Treatment Plants Count", 8 June, http://www.newagebd.com/2009/jun/08/edit.html, accessed 20 December 2009.

PALEN, JOHN A. (1999) "Objectivity as Independence: creating the Society of Environmental Journalists, 1989–1997", *Science Communication* 21(2), pp. 156–71.

PROTHOM ALO (2010) "Freedom of Mass Media and Accountability, a *Prothom Alo* Round Table Discussion", 3 November, http://www.prothom-alo.com/detail/date/2010-11-03/news/106306, accessed 4 November 2010.

RAMAPRASAD, JYOTIKA and RAHMAN, SHAFIQUR (2006) "Tradition with a Twist: a survey of Bangladeshi journalists", *The International Communication Gazette* 68(2), pp. 148–65.

RAHMAN, MATIUR (2005) "When Media Takes a Stand for Social Change: the experience of Prothom Alo in Bangladesh", 5 September, http://www.rmaf.org.ph/Awardees/Lecture/LectureRahmanMat.htm, accessed 22 November 2010.

SCHUDSON, MICHAEL (1978) *Discovering the News*, New York: Basic Books.

SCHUDSON, MICHAEL and ANDERSON, CHRIS (2009) "Objectivity, Professionalism, and Truth Seeking in Journalism", in: Karin Wahl-Jorgensen and Thomas Hanitzsch (Eds), *The Handbook of Journalism Studies*, New York: Routledge, pp. 88–101.

SCHWARTZ, DEBRA A. (2006) *Writing Green, Baltimore*, MD: Apprentice House.

SHANAHAN, JAMES and MCCOMAS, KATHERINE (1997) "Television Portrayal of Environment: 1991–1995", *Journalism & Mass Communication Quarterly* 74, pp. 147–59.

SHARMA, KALPANA (2010) "Good Journalism, That's All", in: Keya Acharya and Frederick Noronha (Eds), *The Green Ink: environmental journalism in India and South Asia*, New Delhi: Sage, pp. 59–63.

SPELLERBERG, IAN S., BUCHAN, GRAEME D. and EARLY, NICK (2006) "Television and Environmental Sustainability: arguing a case for a code of standard in NZ", *Pacific Journalism Review* 12(2), pp. 137–46.

TUCHMAN, GAYE (1978) *Making News: a study in the construction of reality*, New York: Free Press.

WAISBORD, SILVIO (2009) "Advocacy Journalism in a Global Context", in: Karin Wahl-Jorgensen and Thomas Hanitzsch (Eds), *The Handbook of Journalism Studies*, New York: Routledge, pp. 371–84.

WAISBORD, SILVIO and PERUZZOTTI, ENRIQUE (2009) "The Environmental Story That Wasn't: advocacy, journalism and the asambleismo movement in Argentina", *Media, Culture & Society* 31(5), pp. 691–709.

WORLD NEWSPAPER ASSOCIATION (2010) "World Press Trends", www.wan-press.org/worldpres-strends, accessed 30 October 2011.

WYSS, BOB (2008) *Covering the Environment*, New York: Routledge.

PLAYING THE MEDIA GAME

The relative (in)visibility of coal industry interests in media reporting of coal as a climate change issue in Australia

Wendy Bacon and **Chris Nash**

The coal industry is a high-profile and significant sector of the Australian economy that is coming under increasing criticism as a significant source of greenhouse gas emissions. This paper offers a Bourdieusian field analysis of major Australian newspaper coverage of the coal industry in relation to economic, political and environmental issues. Theoretically, it extends Bourdieu's field model to embrace the issue of visibility and invisibility of agents and their positions in the field, using the work of Goffman, Giddens and Ericson. Empirically it finds that both journalists and their sources actively engage with factors of invisibility and visibility in contesting the representation of coal industry interests in media coverage. Overall there is a common journalistic endorsement and visibility across all newspapers of the national and regional economic benefits of the coal mining and power generation industries, but marked differences among individual newspapers in how they position themselves and their sources with respect to climate change impacts and issues. A key factor in the contestation over the journalistic coverage is the production of visibility and invisibility for specific information and activity.

Introduction

A significant amount of international research has contributed to an understanding of the patterns and contours of the media coverage of climate change, demonstrating differences and similarities between national approaches, between publications and across time (Boykoff, 2008; Boykoff and Boykoff, 2007; Carvalho, 2007; Cottle, 2008; Das et al., 2009; Eide et al., 2010).

In surveying the field of media research about climate change, Cottle suggests:

> research into media and the environment has to shift its sights beyond media centrism and the exclusive focus on media practices and processes to examine the interactions between different institutional arenas and how these influence the discourses and representations of environmental news. (2008, p. 77)

In this he follows Schlesinger (1990), but goes on to suggest that an examination of how different "social problems" compete for media visibility may be helpful (following Thompson, 1995, 2000, 2005), and like others he focuses on this issue more from the point of view of environmental groups (e.g. Lester and Hutchins, 2009).

We take a somewhat different approach to questions of visibility, but we agree with Cottle that research into media and climate change needs to complement forms of discourse analysis with other forms of relational analysis that can further elucidate patterns and instances of media coverage in relation to environmental politics, policy formation and decision-making. As a contribution towards this, we will explore the possibilities of a Bourdieusian field analysis of media coverage of the Australian coal industry during a period in which climate change politics are being intensely played out in Australia.

In this paper we are making two main arguments: firstly, that it is impossible to understand the way that coal industry members act to secure their political and economic interests without considering their relationship to journalists and their activities in the media field, and vice versa; and secondly, that an important dimension of the media coverage is the relative visibility and invisibility of sources and their interests in both particular stories and particular categories of coverage.

Coal in Australia

Global production of coal has increased by 38 per cent over the last 20 years. Australia is the fourth biggest coal producer in the world after China, the United States and India, with 121 coal mines in operation currently, concentrated in New South Wales and Queensland. Exploration for new deposits is continual and expanding. Australia is the world's largest coal exporter with currently 46 per cent going to Japan, 37 per cent to the rest of Asia (China's proportion of that trade is increasing) and 12 per cent to Europe (Australian Coal Association, 2009; World Coal Institute, 2009).

Coal production accounts for 2.5 per cent of Australia's gross national product, which would be a surprisingly small proportion to many observers given the high media profile of the industry. However, coal is Australia's number one export income earner, expanding from 19 per cent and A$25 billion in 2005–6 to around A$55 billion dollars in 2008–9. In the media the commodities trade to China, including coal, natural gas, iron ore and other minerals, is generally credited as the cause of Australia's technical (at least) avoidance of a recession during the 2007 onset of the Global Financial Crisis. All Queensland ports reported increases in coal transshipments in 2008–9. Federal and state governments are spending more than $10 billion on port and rail infrastructure in a bid to double Australia's coal exports by 2030 (O'Neill, 2009).

Over 80 per cent of power generation in Australia is currently derived from coal, and coal produces 75 per cent of Australia's greenhouse gas emissions from energy (including transport and heat). Coal mining and distribution is capital intensive and employs only 131,000 people nationally, fewer than the fast food retailer McDonalds. Even in the Latrobe Valley in Victoria, where the brown coal is mined and burned that produces 88 per cent of Victoria's electricity, only 7 per cent of the workforce is involved, less than the 11 per cent employed in the retail industry (Latrobe Council, 2006).

Government revenues in the States of New South Wales and Queensland are heavily dependent on royalties from coal mining. Coal industry interests are influential in the major political parties, all of which acknowledge receipt of substantial financial donations from mining companies. The conservative Opposition parties are strongly aligned with the mining industry, while on the (nationally) governing Labor Party side of politics, major trades unions are aligned both for and against the industry, and those alignments cross traditional left–right boundaries. The Greens, on the other hand, have been advocating a "No new coal" policy and their vote has been increasing substantially at recent elections, to the point where they received nearly 12 per cent of the national vote and held the balance of power in the federal parliament after the August 2010 elections. Clearly coal is at the centre of contemporary politics and economic policy in Australia, and receives significant attention from journalists. The media coverage responds to changes and development in the economic and political spheres as the various institutions and forces contend. Conversely, those institutions and forces respond to the media representations of the issues and developments as they seek to advance or protect their own interests.

Theoretical Framework

Our focus is on the media representation of this contest, but we follow Schlesinger (1990) and Cottle (2008) in arguing that to understand media activity and power they have to be seen in relation to other fields of social power. In this paper we are deploying a field analysis derived from Bourdieu's conceptualisation (Bourdieu and Wacquant, 1992; Benson and Neveu, 2005) to engage with the relations between the economic, political and symbolic (media) fields. Our specific focus and contribution theoretically is on the fact and terms of visibility of sources and their interests as they engage in contestation.

Bourdieu argues that all human social activity takes place within fields, which are socially constructed and contingent on the social relations that construct them and related fields. Any field is a "structuring structure" of the power relations that constitute it, and therefore is necessarily dynamic in time and space. It follows that the boundaries or parameters of a field are determined by the activities of the players in the field; indeed, "the question of the limits of the field . . . is *always at stake in the field itself*, and therefore admits of no *a priori* answer" (Bourdieu and Wacquant, 1992, p. 100, emphasis in original).

Participants (which may be individuals or organisations) take positions in a field defined by their structural or "objective" relations to the stakes of their engagement in that field, including the varying amounts of capital they hold derived from experience, education, resources and skill that they bring to the contest that constitutes that field. The manifestation of capital in the game is termed "habitus". Habitus is a system of dispositions based on the capital the participant has accumulated: it involves the capacity to read and interpret the state of play in the field, the power relations that generated the developments up to that point, and the potential developments that may ensue dependent on the player's interaction with the contest. It is the key element in the "logic or political economy of practice" that links the actions of individuals into the operations of large fields of power relations and that both enables the reproduction of those fields and at the same time their development to cope with changing circumstances.

Much of the interpretation of Bourdieu's work is engaged with the issue of class and other axes of social advantage/disadvantage, in which individuals' and groups' habitus is readily "readable" to other participants in the field and therefore marks the status of the capital one brings to the field. However, it is clear that some degree of dissembling, dissimulation and deception may be advantageous in disguising the parameters of the field, the participation of certain players, the capital they are deploying and their habitus. It is this question of the fact (or not) and terms of participants' visibility in the field that we want to explore in some detail.

A major contemporary theorist of visibility is Thompson (1995, 2000, 2005), who assumes that in a mediated world visibility is a necessary precondition for successful participation in the ideological or symbolic field. However, we argue that if we take a decentred approach to the analysis of media power, and see it in relation to the goals, strategy and actions of players in their major fields of engagement, e.g. the economic or political field, then absolute or relative invisibility may be the most efficacious condition. So invisibility can be a desired tactical and strategic goal in the media field.

Significant theorists of invisibility are Goffman (1969), Giddens (1984) and Ericson (Ericson et al., 1989) who successively refined a theorisation of regions and performance in social life, and all adopted a relational approach to social analysis that is compatible with Bourdieu's relationalism. Goffman (1969) was concerned with the active self-presentation by individuals in everyday contexts such as the workplace. He argued that such activity took place in regions:

> A region may be defined as any place that is bounded to some degree by barriers to perception. Regions vary, of course, in the degree to which they are bounded and according to the media of communication in which the barriers to perception occur. (Goffman, 1969, p. 92)

Using a theatrical metaphor, Goffman argued for a classification of two regions—front and back of house:

> The performance of an individual in a front region may be seen as an effort to give the appearance that his [*sic*] activity in the region maintains and embodies certain standards. (1969, p. 93)
>
> whereas
>
> A back region or backstage may be defined as a place, relative to a given performance, where the impression fostered by the performance is knowingly contradicted as a matter of course. (Goffman, 1969, p. 97)

Giddens took up Goffman's approach to regionalisation of social life, but identified

> limitations of the dramaturgical model that Goffman employs ... If agents are only players on a stage, hiding their true selves behind the masks they assume for the occasion, the social world would indeed be largely empty of substance ... The sustaining of ontological security could not be achieved if front regions were no more than facades. (Giddens, 1984, p. 125)

We can note the importance of Goffman's insight about performance and deception in social life, but to accommodate the need for revelatory self-representation as well as the deceptive self-representation of Goffman, Giddens proposed an intersecting axis of enclosure/disclosure to complement the front/back axis (1984, p. 124). Ericson developed Giddens' model further by extending its application from individuals to organisations in their dealings with the news media, and argued that it was a major role of communications staff in dealing with journalists to "police the boundaries" between the four regions to maintain their integrity and security (Ericson et al., 1989, p. 9).

There is a normative element to Ericson's spatial model (Figure 1), in that the enclosed/front of house region (top left quarter) amounts to failure to disclose, suppression or even censorship of what should be on the public record, as distinct from the enclosed back region which is legitimately the province of organisational privacy and secrecy. Journalists, and in particular investigative journalists, probe all these regional

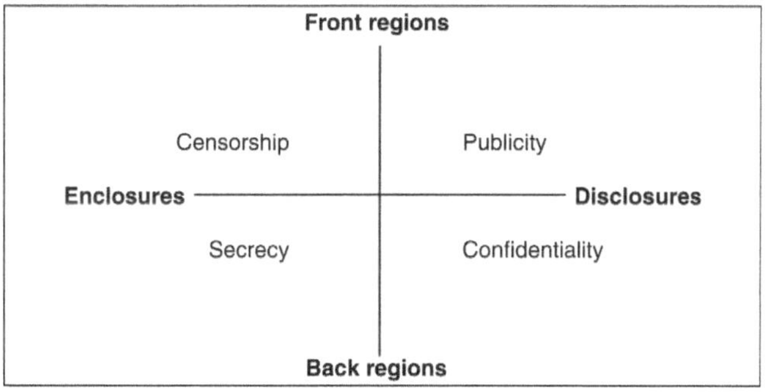

FIGURE 1
Ericson's spatial model. Source: Ericson et al. (1989, p. 9)

barriers to bring information into public visibility, and organisations dissemble and misinform journalists as they play the politics of the media field to secure their organisational goals.

In these encounters with their sources within the field, journalists can adopt a range of editorial postures, e.g. news, extended analysis, opinion, investigation and exposé. There are varying conventions related to the use and display of the journalist's structural position and habitus in these diverse genres, and Ericson tends to underplay the significance of journalists' flexible positioning within the field, preferring to focus on the activities of organisations as the generative source of news information (Ericson et al., 1989, p. 1). Similarly, he does not deal with journalists' extensive use of "off the record" sources and the role of journalists as investigators and revealers of hidden or enclosed information.

Every development of a story by a journalist involves a negotiation (explicit and implicit) with sources about what is relevant from the contemporary and socio-historical contexts (whether proximate or more removed) and what can be safely ignored and quickly/eventually forgotten. Ericson does not deploy a temporal dimension to his spatial model, though he does mention in passing the tactical issues of timing (Ericson et al., 1989, p. 20); however, the rhythmic and arhythmic temporalities of the news cycle can structure the visibility, invisibility or obscurity of information and activities at least as effectively as the spatial characteristics of access. Ericson also explicitly downplays the relevance of larger social power relations in his research (Ericson et al., 1989, p. 3), and so unsurprisingly his case studies suggest that there are no larger power relations structuring the relations between journalists and their sources. Bourdieu's theorisation of the relationship between fields, and his concern with the links between "subjective" activity by agents and the broader social "objective" structure of fields remedies these lacunae in Ericson's approach.

To summarise our theoretical approach, we argue that a dynamic field analysis that examines journalistic reportage in relationship to the fields of interest and activity of source organisations is a productive approach to researching the contestation of media power in the representation of issues. A significant, under-researched dimension of field analyses is the terms of relative visibility/invisibility of information in time and space, and the Ericson model provides a valuable extension of Bourdieu's framework in this regard.

The Media Coverage

Empirically, our analysis covers a month-long period in July 2009 of mentions of "coal" in three major daily newspapers in New South Wales and Queensland to establish the bedrock scope and structure of daily reporting of coal; that is to say, to identify the structured positions adopted by journalists and sources. We selected the month of July 2009 for several reasons.

Firstly, it was a period of intense negotiation between the Labor government and coal industry lobbyists over the disbursement of "free permits" for greenhouse gas emissions under the proposed Emissions Trading Scheme (ETS) that had been passed by the House of Representatives but was being delayed in the Senate then under the control of the Opposition parties.

Secondly, the looming deadline for the Senate to pass the bill was 14 August, and there was intense contestation within the Opposition parties over whether the bill should be passed, contestation that resulted in a change of leadership in the Liberal Party and the adoption of what was effectively a climate change sceptic position by the Opposition.

Thirdly, the Global Financial Crisis was having a significant impact on the domestic Australian economy, but it became clear during July 2009 that exports were rising again, especially of coal to China. Mining companies announced strong yearly results and mine expansions, and the commodities trade with China was widely identified by politicians, business and the media as the factor that might enable Australia to avoid an economic recession.

Fourthly, during this month the (then) Australian Prime Minister Kevin Rudd attended the G8 conference in Rome where, at US President Obama's invitation, he announced the establishment in Australia of the Global Carbon Capture and Storage Institute as the vehicle for the international research and development effort for "clean coal".

Fifthly, there was a major corruption trial under way in Queensland involving bribes paid by a coal company to a politician, revealing hidden aspects of the political–economic relationship. Collectively, these five factors generated a scale and scope to the journalistic reporting of coal that delineated the structure of the interrelated fields and the interests and activities of the agents (sources and journalists) deployed within them.

A database search was conducted for the word "coal" in major newspapers on the eastern seaboard of Australia in the coal-rich states of NSW and Queensland: Sydney's two metropolitan newspapers—*The Sydney Morning Herald* (*SMH*), a broadsheet owned by Fairfax Media with a daily circulation of over 200,000 targeted at higher socio-economic status readers particularly in the wealthier areas of Sydney, and the tabloid *The Daily Telegraph* (*DT*) owned by Rupert Murdoch's News Corporation with approximately 350,000 readers targeted at a more conservative and not so wealthy audience, along with *The Courier Mail* (*CM*), the only metropolitan daily newspaper in the Queensland capital of Brisbane, also owned by News Corporation and with a similar circulation to *The Sydney Morning Herald*.

We coded the stories into topic themes identified on a generative basis from the stories. We eliminated several stories where the reference to coal was a passing one in the context of travel or history stories. The rest of the stories were grouped under topic themes:

- emphasis on climate change generally (including the ETS);
- focus on "clean coal";
- coal in relation to other energy sources, e.g. nuclear, wind, solar energy;
- environmental and other problems associated with the coal industry apart from climate change;

- corruption and other improper practices associated with the coal industry;
- business, investment and expansion of the coal industry;
- export of coal;
- allied industries with coal—transport, commercial property, banking, etc.

Table 1 quantifies the results, which demonstrate that the profiles of coverage of coal by the newspapers diverged considerably. *CM* had by far the largest amount of coverage, though that was skewed by the coverage of the trial and conviction of former Queensland Labor Government Minister Gordon Nuttall for receiving bribes from Ken Talbot, then CEO of Macarthur Coal, a major coal company. In contrast, there was only one report of this trial in Sydney, in the *SMH*, which also carried a report of another coal corruption story, but there was no mention of the trial at all in the *DT*.

Queensland's *CM* was the most positive newspaper in its coverage of the coal industry, with 43 articles about its economic benefits and only 25 about climate change policies, politics and other problems associated with coal. This compared to an inverse polarity in the *SMH* of 22 business-related stories and 32 climate change and environment-related stories. Even the *DT*, a large circulation Sydney tabloid with a generally right-wing orientation, published 13 stories about climate change and other environmental issues versus seven about economic benefits and importance, and although the articles about climate change did not necessarily endorse the concept, nonetheless they did acknowledge the issue and engage with it, hence confirming its visibility in the field. Each of the

TABLE 1
Reporting of coal stories by thematic topic and masthead

Topic/issue	*The Sydney Morning Herald*	*The Daily Telegraph*	*The Courier Mail*
Business			
Business investment or expansion in coal	15	3	17
Trade (especially in relation to China/Japan), Global Financial Crisis, employment	5	4	10
Allied industries: infrastructure investment, real estate, privatisation	2	0	16
Subtotal	**22**	**7**	**43**
Environment			
Climate change politics: Emissions Trading Scheme, policy, protest	18	3	14
Clean coal	2	2	3
Alternative energy policy/politics: nuclear, wind, etc. (mentions coal)	6	3	4
Non-greenhouse gas emission problems with coal—taking land from farming, dams, safety, marine damage and pollution, offshore drilling threat	6	5	4
Subtotal	**32**	**13**	**25**
Corruption/improper practices			
Political corruption/secret commissions involving people with coal interests	2	0	26
Subtotal	**2**	**0**	**26**
Total	**56**	**20**	**94**

three mastheads also reported environmental problems unrelated to climate change caused by the coal industry, at a rate compared to the economic reports of 41 per cent at the *DT*, 27 per cent at the *SMH* but only 9 per cent at the *CM*.

Almost all business stories across all mastheads were based on an underlying assumption and perspective that expansion of the coal industry was positive for Australia. Much of this news came in the form of corporate reports supplied by the Australian Stock Exchange or the Reserve Bank that were reported positively with little or no further discussion or analysis. They occupy the front/disclosure region (top right quarter) of Ericson's chart (Figure 1), occasionally supplemented by confidential information (bottom right quarter) embellishing the story. In their business reporting the media in the symbolic field are taking a structural position that is tightly aligned with the interests of the coal industry in the economic field. It represents and endorses the dominant "business as usual" structure of the economic field within which the contest over climate change policy is conducted. Further, the business reporting by and large does not engage with the climate change issue, which is invisible in these stories. Indeed, there was very little discursive interaction between the two themes in the coverage: business stories by and large did not engage with environmental issues, and vice versa. The two issues were effectively sub-fields of the coverage of coal, operating relatively independently and effectively offering competing perspectives as to the significance and meaning of the coal industry, and leaving it to readers to draw their own conclusions on any putative links between the two. Certainly none of the newspapers could be said to be trying to reconcile these two perspectives on the issue of coal in the symbolic field.

While they had a common approach to reporting on the economic dimension of the coal industry, the newspapers sharply diverged from each other in their coverage of the environmental dimension of the coal industry, and we will explore these divergences in some detail, to depict and analyse the cut and thrust of contestation and the differing relationships of the newspapers to the unfolding struggles.

The Daily Telegraph on Climate Change

As indicated above, July 2009 was a period of intense controversy over the government's proposed ETS. While the *SMH* and the *CM* both engaged intensively with the controversy during July 2009, albeit from different points of view, the *DT* dismissed it in two strident opinion pieces and a single news report. The first opinion piece was written by outspoken right-wing columnist Piers Akerman, who characterised the science of climate change as widely contested and supported the views of climate sceptic Dr Garth Paltridge. Akerman deployed a rhetoric of revelation and free debate confronting suppression of dissenting viewpoints, but failed to make visible for readers the lobbyist connections and industry alignment of the climate sceptic he promoted, effectively equating scientific and industry perspectives. In this he was typical of media promotions of the sceptic position (Boykoff and Boykoff, 2007; Chubb and Nash, 2010; Chubb et al., 2009).

The second opinion piece in the *DT* took a rather different approach, not contesting the science of climate change but challenging the ETS as an appropriate response. It was authored by News Ltd economics editor Terry McCann, also outspokenly right-wing in his views. McCrann asserted that the G8 conference attended by Rudd and Obama in Rome demonstrated that the "climate change game is over". Because China had failed to demonstrate a commitment to curbing emissions, "under no circumstances should we lock in any sort of emissions trading scheme... [t]he ETS is national suicide anyway. Emitting carbon dioxide is this country's core and irreplaceable comparative economic advantage" ("Emissions Talk Just a Lot of Hot Air", *DT*, 15 July 2009). McCrann's article was republished in *The Courier Mail.*

This coverage in the *DT* sought to cover both sides of the argument: there is no climate change problem, but even if there were, the ETS is not the answer and will do irreparable damage to the Australian economy. The same arguments were being mounted at the time by the members of the Opposition parties opposed to the ETS, and who were shortly to triumph in deposing ETS supporter Malcolm Turnbull from the Opposition leadership. The *DT* was effectively aligning itself in the symbolic field with the sceptics in the Opposition parties in the political field on the ETS issue, and this partisan alignment was steady, visible and noted by commentators in competing media in the period up to and beyond the August 2010 national elections.

The *DT* did publish a report about the damage that offshore queues of up to 40 coal ships could do to marine life when they break or drag their chains ("Lost Chain Litters Sea", *DT*, 7 July 2009) and several environmental stories about the threat of planned off-shore exploration for coal to whales and the whale-watching tourism industry. On the same day, the *DT* carried an editorial arguing that the "barbaric" proposal "sounds surplus to requirements. Given the already vast reserves in NSW land areas, one wonders why exploration should be seriously considered at this point" ("Mining Threat to Whales", *DT*, 8 July 2009). This report and editorial provide an illuminating example of the intricacies of media contestation over visibility. The protection of whales and the whale-watching industry are an unchallenged public good in Australian media and public discourse. The threat to whales from offshore coal exploration was therefore an eminently publishable story enabling the *DT* to align itself editorially as the public voice of its readership on this issue. The conflict with the coal industry interests was managed by distinguishing between onshore coal mining (which is to be approved) and offshore coal mining (which is to be disapproved).

In response, the head of the company that had applied to the Federal government for a licence was quoted as saying: "I have no idea where the humpback whales migrate... but perhaps the councillors would not like to drive around in cars or live in nice warm houses. You can't have your cake and eat it too. It is unrealistic to say No Mining" ("Mining Threat to Whales", *DT*, 8 July 2009). The following day, perhaps trying to repair the damage from such a clumsy response, the public company which owns the project put out a stock exchange announcement that they were in fact researching the impacts of their plans on whales, which in any case were being much more threatened by the coal ship traffic ("East Coast Minerals Examining Whale Migration", ASX announcement, 9 July 2009). That release, directed as it was to business reporters who were not following the story, was not reported, and remained invisible.

The Sydney Morning Herald on Climate Change

During 2009, the *SMH* placed a high priority on the coverage of climate change politics, and accorded a stronger critical emphasis to Australia's actions in the international context than either the *DT* or the *CM*. The Environmental Editor Marion Wilkinson, one of Australia's most experienced investigative and political journalists, began the month with an analytical piece summing up the international and national situation that ended with the suggestion that Australian politicians who did not support a stronger response to climate change would be seen by future generations as no better than climate sceptics. This article clearly promotes action in the political and economic fields to address climate change issues ("Fatal Flaw in Turnbull's Climate Compromise", *SMH*, 2 July 2009).

In addition to such campaigning articles from its own environmental reporters, the *SMH* republished a strong column by *New York Times* columnist Thomas Friedman arguing that the US economy would be left behind by China if it did not prioritise renewable energy ("China Tests Wind on Clean Energy and Makes Its Move", *SMH*, 11 July 2009). Later in the month, the *SMH* published an opinion piece by Climate Project India director Gaurav Gupta, which ended:

> When Rudd says "we are all in this together", he glosses over the deep inequities of climate change. Australians have enjoyed the benefits of more than a century of untaxed carbon emissions floating into the global atmosphere. ("The Mucky Country: we outsource our pollution", *SMH*, 23 July 2009)

These two columns reflect the more international focus in the *SMH* than in the other two papers, and again demand action in the political and economic fields by making visible the international context of Australia's position.

In another story, Wilkinson reported on NSW power generation industry documents that were publicly available but published for the first time by her. According to the reports, rather than planning to reduce emissions the power companies were aiming to expand their coal and gas energy output. She backgrounded the companies' actions in the context of NSW already having two of the ten most polluting electricity plants in Australia and compared them unfavourably to US plans to block new coal plants ("State's High-polluting Power Plants Seek to Expand Their Output", *SMH*, 22 July 2009). The *SMH* also carried stories that critiqued the concept of "clean coal", and accused the NSW Resources Minister of corruptly giving a coal exploration licence to a friend and allowing valuable farmland to be mined.

By way of balance, the NSW parliamentary reporter produced a report about documents warning of "severe economic dislocation as a result of moves to slash Greenhouse emissions" that had been tabled in the NSW Parliament ("Mine Districts Will Pay Dearly for Cuts: report", 11 July 2009). This report did mention that the documents were based on advice by consultancy firm ACIL Tasman but did not disclose that the firm had a long record acting on behalf of fossil fuel interests promoting climate scepticism (Pearse, 2007, p. 305).

The complexity of the contestation over visibility is well-illustrated by the *SMH* report on a speech by the Secretary of the Federal Department of Climate Change, Dr Martin Parkinson, in which he had "blasted" the mining sector and lobbyists and "lashed out" at their models that forecast thousands of job losses in areas such as coal. "Too often these reports are driven by a pre-determined headline. Too often they contain low quality analysis, flawed assumptions and selected use of data", he said ("Call for Honesty in Emissions Debate", *SMH*, 24 July 2009). Public servants in Australia require specific permission to be interviewed by journalists "on the record", that is, for attribution as sources. This leads to a general neglect of reporting on the important administrative field of government, and forces journalists to rely on insider, undisclosed or "off the record" sources (Ericson's back/disclosure region—the bottom right quarter of his chart). Government communications officers, as Ericson puts it, "patrol the boundaries" (Ericson et al., 1989, p. 9). The *SMH* reported that Parkinson's speech was very political for an Australian public servant and that it "believed" the Government (i.e. politicians) had "considerable input into its writing". The paper withdrew this allegation the following day after Parkinson objected to this description. It is likely that someone close to Parkinson had given the paper this information but was forced to retract it. This story also reported that "inside negotiations" between business and the conservative Opposition were being discussed in case it decided to change its position and vote for the ETS scheme.

By the end of the month, the coal sector were deploying their economic capital to bypass journalists and intervene directly by placing political advertisements in the media warning of job losses if the ETS was passed, which may be interpreted as a sign of industry dissatisfaction with the terms of the visibility of their position on climate change and a decision to appeal directly to the readership. Meanwhile the *SMH* quoted "unnamed government experts" as saying that to give in to the coal industry would cost the scheme $A10 billion and force the government to cut compensation to ordinary households. "Senior departmental officials" were reported as "urging the government to stand firm" and that they were "privately disputing" the claims of the Opposition and the coal industry. The *SMH* was probing the "back regions" of the political field to play an active role in shaping and moving the political debate. As later became apparent, their tactics were placing the communications section of the Department of Climate Change under pressure. In early 2010, Sydney's *Sunday Telegraph* reported on a "leaked memo" by Department head, Dr Parkinson announcing:

> Given the heightened focus on climate-change communication, I have also decided to bolster the resources devoted to this task . . . [t]he communications branch is a critical whole-of-department resource. The department faces a challenging year ahead and the changes outlined in this notice are designed to strengthen our capacity to meet these challenges and Government objectives. ("A Change of Climate Spinners", *Sunday Telegraph*, 21 February 2010)

The following month, under questioning in a Senate committee, Dr Parkinson admitted that he learned of the Rudd Labor government's April 2010 shock decision to drop the ETS bill only when it was leaked to the media. He said the decision had left many officials "terribly, terribly disappointed" and told the Committee that staff were establishing "time capsules" of their work to be opened "whenever this current impasse is broken and we can have an appropriate debate . . . around climate change" ("Officials Discovered ETS Was Canned in the Media", *The Age*, 28 May 2010). Players in the game from all sides use the media to communicate with their own members as well as with the public, and such reports make visible the tenor of internal departmental relations.

A theme in the *SMH* coverage of climate change was that fossil fuel interests including the coal industry were seeking to exert pressure on government decision-making through unpublicised lobbying activities and electoral donations. As far back as 2005, the *SMH*'s sister paper, *The Age* in Melbourne, had used Freedom of Information searches and other investigative techniques to reveal how BHP Billiton, Rio Tinto and other big polluters were influencing government policy ("How Big Energy Won the Climate Debate", *The Age*, 30 July 2005).

Although only published subsequently, in July 2009 the *SMH* environmental team were already working on their contribution to an International Consortium of Investigative Journalists' investigation of the fossil fuel lobby. Shortly before COP 15 in late 2009, reports were published in United States, Canada, Australia and Belgium documenting the links of the fossil fuel industry to lobbying activities. The *SMH* series began:

> Big greenhouse polluting companies around the world, employing thousands of lobbyists, are exerting heavy pressure on governments to weaken climate change laws at home and slow progress on an international climate agreement in Copenhagen, a global investigation reveals. In Australia, 20 companies who have already won the most concessions from the Rudd Government's emissions trading scheme employ 28 lobbying firms with well over 100 staff, many of them former politicians, political advisers or government officials . . . They include John Dawkins, a former federal treasurer who is now a director of lobbying firm Government Relations Australia that represents BP, Peabody Resources, and The National Generators Forum, a power industry trade group. ("Come in Spinner", *SMH*, 11 July 2009)

This investigation followed on from the widely publicised 2007 book of Guy Pearse, which exposed the significant role industry lobbyists played in informing and often actually devising government policy (Pearse, 2007). The perception that lobbyists have been exerting their influence in unacceptable ways is a politically potent notion that highlights the issue of invisibility. This mode of crusading investigative journalism accesses documents and insider accounts to make visible the enclosed back regions of players in the economic and political fields. Apart from the presentation of undisclosed and secret information for public consideration, the very act of breaching the barriers to entry has an impact on the players in those fields, requiring them to re-group to secure their organisational frontiers and deploy further resources to contest the symbolic field, as demonstrated in the Parkinson/public servants example above. Processes in the symbolic, economic and political fields become actively intertwined by the act of making information visible against the wishes of the information holders.

The Courier Mail on Climate Change

Like the *SMH*, *CM* focused strongly on the ETS scheme but its approach was conditioned by its perception of its own role as an active and interventionist defender of the interests of Queensland, where it is the only major newspaper. It interpreted Queensland's interests as primarily economic in nature and gave strong visibility to economic stories on coal, publishing twice as many articles (43) over the month as the *SMH* (22) on that topic, compared to 25 on environmental issues (32 in the *SMH*). The *CM* also gave almost daily coverage (26 articles) to the corruption trial involving former Minister for Infrastructure Gordon Nuttall and Macarthur Coal CEO Ken Talbot.

The *CM* began the month with a critical column by a staff reporter alleging that the Queensland Labor government believed that its coal industry was being "dudded" by the Federal Labor government's ETS scheme but that Queensland Premier Anna Bligh was "barely saying so publicly" ("Hot Words", *CM*, 4 July 2009).

The *CM* admonition of Anna Bligh seemed to bear results when she came out publicly calling on her Federal colleagues to increase the compensation for the coal industry. On 16 July, a *CM* story led with the words: "Queensland's coal industry has praised Premier Anna Bligh's decision to 'go into bat' for it by trying to secure taxpayer money for coalmines under a Federal plan to slash greenhouse gas." The coal industry lobby group AIGN was quoted as welcoming Bligh's comments, which were also supported by another industry peak organisation, the Queensland Resources Council. Although positive to both the coal interests and Bligh, the report was balanced by quotes from a World Wildlife Fund spokesperson which pointed out the coal industry was already receiving $750 million compensation under the scheme ("Coal Industry Backs Bligh Bid to Raise ETS Compo", *CM*, 16 July 2009).

A further story on the same day reported that the Federal MP for the electorate that included the Queensland coal port of Gladstone wanted his government to "do better" to compensate the coal industry. The Queensland Resource Council was again quoted, referring to a report by consultancy group Access Economics' (a regular protagonist arguing the case against action on climate change; Pearse, 2007, pp. 157, 208) finding that the ETS scheme could cost Queensland 28,000 jobs. Finally, the Leader of the Opposition was quoted as also demanding more Rudd government support for the coal industry ("Federal MP Joins Chorus—emissions payout is 'too little'", *CM*, 16 July 2009). The impact was further strengthened by an editorial the following day arguing that although the *CM* remained in favour of some sort of eventual action, the current ETS bill should be delayed until the "rest of the world" was prepared to "follow our example" ("Caution Vital in Emissions Trading Plan", *CM*, 17 July 2009).

This suite of stories encapsulated the *CM* position in reporting coal: strong reporting and visibility for industry sources on economic claims, buttressed by opinion pieces and editorials supporting industry expansion and compensation in the event of an ETS being enacted. For *The Courier-Mail*, supporting Queensland equated to supporting the coal industry, and it maintained pressure on the State Labor government to do so, even if necessary against the Federal Labor government.

The *CM* also carried stories that were positive about "clean coal" as a possible technological solution to carbon emissions, which again contrasted with the more critical coverage of this issue in the *SMH* and the *DT*. It published a very positive report of how US President Barack Obama had shared the stage at the G8 meeting in Rome with Prime Minister Rudd who announced Australia's $100 million a year Global Capture and Storage Institute backed by "23 governments and more than 100 companies" ("Green Power Allies—Obama and Rudd revisit a meeting of minds moment", *CM*, 11 July 2009). It followed this article two weeks later with an upbeat report that "An international consortium wants to build a $3 billion-plus 'clean coal' power plant at Wandoan in Queensland's Surat Basin. It would use technology that converts coal into a cleaner burning fuel and captures and stores carbon dioxide emissions." An executive of GE, which is investing in the project, was quoted as saying, "What is really exciting is this opportunity to deploy IGCC technology with CCS (carbon capture and storage) and deliver low carbon coal power in Australia" ("New Coal Plant at Centre of Global Bid", *CM*, 27 July 2009).

During the month, *The Courier Mail* carried many court reports of the corruption trial of Gordon Nuttall who served as Minister in several portfolios while he was receiving bribes from one of Queensland's richest men, the ex-CEO of major coal company Macarthur Coal, Ken Talbot. Nuttall was convicted later in 2009, but the advantage that Talbot achieved from the bribes was yet to be established. Talbot was to be tried in 2010 but was killed in a plane crash while on a business trip in Africa. In the light of Australian laws covering contempt of court, it is not surprising that there was no broader discussion about the underlying issues during 2009. In this case, the *CM* restricted its coverage of the corruption issues to a report of Talbot's defence as explained to them in an unpublished interview before he died, and apart from that one instance, the coverage of Talbot celebrated his contribution to Queensland and the coal industry. Thus the detail of the purpose of the bribe remained invisible, for the moment at least.

Discussion

We can make a number of observations on this study. Firstly, it is not possible to understand the shape or contours of media coverage without reference to the specificities of relations and interests in the fields being reported, in this case the political and economic fields of Queensland and NSW, the contest in the Opposition Coalition parties about the ETS, the coal industry wanting more compensation under the proposed ETS bill, and so on. This relationship between the fields is dynamic and mutually reflexive, because not only are the source organisations trying to influence media coverage, but quite clearly the media are actively playing the game in the symbolic field to influence events in the political and economic fields. This game playing can be seen not only in terms of the subtler forms of representation and selection of sources, however important that may be, but it involved strategic acts of discovery and release of information, demanding policy responses and probing back regions in ways that will influence policy makers.

Secondly, media politics are played out in large part by both the fact and terms of visibility/invisibility of sources in specific reporting contexts. Coal mining and export and coal-fired power generation are presented by sources with a vested business interest as an unalloyed benefit economically, particularly in the context of the Global Financial Crisis, and environmental or other critics of the industry are not visible in those economic stories. Conversely, environmental stories about coal are largely kept quite separate from business stories, and in our month of coverage totalling 170 stories there were only three exceptions: two were opinion pieces in the *DT* (one of which was republished in the *CM*) which predicted severe economic damage from political action to counteract climate change, and a third report in the *SMH* of documents tabled in parliament also predicted "severe economic dislocation".

Thirdly, sources act (deploying their habitus) to manage visibility and police the information boundaries as suggested by Ericson et al. (1989) and this can involve active, though not necessarily successful, dissimulation. Coal industry sources pro-actively pursue publicity to promote their business interests, but are largely passive and reactive in relation to environmental stories. Interestingly, climate change sceptics are barely visible as sources in the business coverage of coal, except in the opinion piece by Akerman. This suggests that business sources no longer see advantage in publicly aligning with denial or sceptic positions and instead prefer to negotiate around the terms of compensation and the timetable for implementation of an ETS or other schemes. Business sources that want to buy into climate policy negotiations have little to gain in advocating the denial position.

Fourthly, in some fields journalists are very active players in seeking to influence events in other fields. They pursue this activity by seeking to make visible information from the enclosed regions of organisational information, for example the public service, thereby forcing organisations to modify their actions in response to public scrutiny. In our study the active investigative research to seek out hidden information was largely done by environmental journalists, while the business and financial reporters largely gave publicity to information voluntarily provided by their sources. The exception to this was *The Courier-Mail*'s examination of the Queensland budget papers to campaign for greater compensation for local businesses. Information can become partly visible through exposure and therefore relationships may shift in response to visibility.

Field analyses are a powerful tool for investigating the interactions between the media and other areas of social activity. They can generate a dynamic and detailed analysis capable of relating the subtleties of individual actions by protagonists in the field to the larger contours of social developments and shifts, which is precisely the aim that Bourdieu set himself in elaborating his approach. Critical to the analysis of power in the media field is the issue of visibility, which must necessarily include the advantages as well as disadvantages of invisibility.

REFERENCES

AUSTRALIAN COAL ASSOCIATION (2009) "The Facts About the Australian Coal Industry", www.australiancoal.com.au/. . ./ACA_facts_sheet_final_2809.pdf, accessed 12 September 2009.

BACON, WENDY and NASH, CHRIS (2006) "Reporting Sustainability in the English-language Press of Southeast Asia", *Pacific Journalism Review* 12(2), pp. 106–35.

BENSON, RODNEY and NEVEU, ERIC (2005) *Bourdieu and the Journalistic Field*, Cambridge: Polity Press.

BOURDIEU, PIERRE and WAQUANT, LUIS (1992) *An Invitation to Reflexive Sociology*, Chicago: University of Chicago Press.

BOYKOFF, MAXWELL T. (2008) "The Cultural Politics of Climate Change Discourse in UK Tabloids", *Political Geography* 27(5), pp. 549–69.

BOYKOFF, MAXWELL T. and BOYKOFF, JULES M. (2007) "Climate Change and Journalistic Norms: a case study of U.S. mass-media coverage", *Geoforum* 38(6), pp. 1190–204.

CARVALHO, ANABELA (2007) "Ideological Cultures and Media Discourses on Scientific Knowledge: re-reading news on climate change", *Public Understanding of Science* 16(2), pp. 223–43.

CHUBB, PHILLIP and NASH, CHRIS (2010) "The Politics of Reporting Climate Change at the Australian Broadcasting Corporation", paper presented to the Communicating Climate Change II Conference, University of Hamburg, October.

CHUBB, PHILLIP, NASH, CHRIS and BIRNBAUER, WILLIAM (2009) "Fighting Over Fires: climate change and the Victorian bushfires of 2009", paper presented to the Global Dialogue Conference, Aarhus University.

COTTLE, SIMON (2008) *Global Crisis Reporting*, London: Open University Press.

DAS, JAHNNABI, BACON, WENDY and ZAMAN, AKHTERUZ (2009) "Covering Environmental Issues and Global Warming in Delta Land: a study of three newspapers", *Pacific Journalism Review* 15(2), pp. 10–33.

EIDE, ELIZABETH, KUNELIUS, RISTO and KUMPU, VILLE (Eds) (2010) *Global Climate–Local Journalisms. A transnational study of how media make sense of climate summits*, Global Journalism Research Series Vol. 3, Bochum: Projekt Verlag.

ERICSON, RICHARD, BARANEK, PATRICIA and CHAN, JANET (1989) *Negotiating Control: a study of news sources*, Milton Keynes: Open University Press.

GIDDENS, ANTHONY (1984) *The Constitution of Society: outline of the theory of structuration*, Cambridge: Polity Press.

GOFFMAN, ERVING (1969) *The Presentation of Self in Everyday Life*, London: Allen Lane, Penguin Press.

LATROBE COUNCIL (2006) "What Industries Do We Work In? Time series industry categorgies", http://profile.id.com.au/Default.aspx?id=294&pg=139&gid=10&type=enum, accessed 2 October 2010.

LESTER, LIBBY and HUTCHINS, BRETT (2009) "Power Games: environmental protest, news media and the internet", *Media, Culture & Society* 31(4), pp. 579–95.

O'NEILL, MARGOT (2009) "Coal Hard Facts of Climate Change Policy", Australian Broadcasting Corporation, 8 October, http://www.abc.net.au/news/stories/2009/12/02/2759398.htm, accessed 30 September 2010.

PEARSE, GUY (2007) *High & Dry: John Howard, climate change and the selling of Australia's Future*, London: Penguin.

SCHLESINGER, PHILIP (1990) "Rethinking the Sociology of Journalism: source strategies and the limits of media-centrism", in: Marjorie Ferguson (Ed.), *Public Communication: the new imperatives: future directions in media research*, London: Sage, pp. 61–83.

THOMPSON, JOHN B. (1995) *The Media and Modernity: a social theory of the media*, Cambridge: Polity Press.

THOMPSON, JOHN B. (2000) *Political Scandal: power and visibility in the media age*, Cambridge: Polity Press.

THOMPSON, JOHN B. (2005) "The New Visibility", *Theory, Culture and Society* 22(6), pp. 31–51.

WORLD COAL ASSOCIATION (2009) "Coal Facts 2009", http://www.worldcoal.org/resources/coal-statistics/, accessed 30 September 2009.

WORLD COAL INSTITUTE (2009) "The Coal Resource", www.worldcoal.org/.../coal_resource_overview_of_coal_report(03_06_2009).pdf, accessed 15 September 2010.

SETTING THE AGENDA ON ENVIRONMENTAL NEWS IN NORWAY
NGOs and newspapers

Roy Krøvel

This paper contributes to the understanding of the developing relationships between non-governmental organisations (NGOs), journalists and environmental issues in general, and climate change in particular. It follows 17 Norwegian NGOs over a period of 10 years, evaluating their ability to set the agenda for public debate on environmental issues in Norway, about the types of organisations that have succeeded in setting the agenda in Norwegian print media over the last 10 years. What role does the emergence of climate change as a dominating global environmental issue play in the agenda-setting capacity of Norwegian NGOs? The results do not support earlier findings that highlight an increasing media focus on sensation and identification and a lack of depth and quality. By contrast, the "winners" found in this investigation are generally those organisations that focus on producing knowledge, and not those that focus on supposedly media-friendly activism. The results also stand in contrast to existing literature on the "success" of Greenpeace. Greenpeace and similar organisations found it hard to compete for media attention as the media interest turned from pollution, roads, dams and local issues to the very complex matter of global climate change. These findings are, finally, used to reflect on how journalists employ journalistic norms.

Introduction

According to Ytterstad and Eide, Norwegian non-governmental organisations (NGOs) wield a significant influence over the Norwegian media in respect of environmental issues, through the process of agenda setting, the framing of news and the selection of sources (Eide and Ytterstad, 2009). The NGOs employ numerous information managers and are becoming increasingly adept at using the media to promote themselves or the issues for which they struggle. It is therefore important to investigate the relationships between the media and NGOs. Changes, over time, in these relationships may help us to understand how journalists work on environmental issues.

This paper contributes to the understanding of the developing interaction between NGOs and journalists around environmental issues in general and climate change in particular. Choosing a global challenge such as climate change encourages reflection on some of the ways in which global issues confront established journalistic practices.

The investigation starts in the late 1990s. At that time, only a handful of studies on journalism, the media and the environment in Norway had been published (Ihlen, 1999, 2001a, 2001b). According to Johnsen in an early contribution (Johnsen, 2006), the general development towards increasing speed in news journalism can have a negative influence on the ability of the audience to understand the implications of complex issues. Johnsen also disapproves of the general style of news presentation, which he sees as a hindrance to more in-depth journalism on the long-term effects of environmental degradation (Johnsen, 1994). Nonetheless, Johnsen observes an increasing interest in environmental questions in Norwegian newsrooms.

Mørck suggests that media reporting of the environment can be divided into three groups: small news items for those with a special interest in the environment; dramatic news on single issues; and poorly camouflaged PR for business or industry (Mørck, 1996).

Investigating the agenda-setting capacity of NGOs requires an analysis of at least three different players and processes. First, because diverse organisations employ different media strategies, variances over time in their visibility in the media might be due to differences—or changes—in these media strategies. Second, different media and different newspapers employ diverse sets of criteria to evaluate the newsworthiness of the information provided by players such as NGOs. Variances, over time, in their visibility in the media could therefore be due to developing criteria or changes in the use of the criteria. Third, the Norwegian media landscape has changed significantly over the last 10 years. Norwegians in general now have access to many more television channels than previously, and depend more on the Internet for information. This has also affected the role newspapers play in setting the agenda for public debate.

This article deals with only a part of this complex whole. The intention is to open up this field for future research by testing some existing hypothesises on NGOs and the media. The investigation follows 17 Norwegian NGOs over a period of 10 years. The article discusses possible explanations for this outcome based on existing literature about Norwegian NGOs and Norwegian media. Finally, it reflects on possible explanations for the findings in the quantitative part, using qualitative methods. Combining agenda setting with qualitative research can often help to avoid some of the criticisms levelled at traditional agenda-setting methodology (Kwansah-Aidoo, 2001).

The research questions for the article are: first, what types of organisations have succeeded in setting the agenda in Norwegian print media over the last 10 years? Second, what role does the emergence of climate change as a dominating global environmental issue play in the agenda-setting capacity of Norwegian NGOs?

Existing Literature on the Media, Environmental NGOs and Climate Change

Over the last 10 years, a large number of NGOs have been competing to set the agenda on the environment in Norway. Distributing information is a core activity for non-governmental organisations, according to Weyker and Willets (Weyker, 2002; Willets, 1982), and NGO activists themselves stress the importance of using the media to inform and influence the public. Information is "the business" of many NGOs (Brown and May, 1991). But according to Weyker and Willets not all information is effective. First, to be effective the information must be perceived as trustworthy or "objective"; if it is not, it will run the risk of being ignored. Second, the information must be produced and distributed fast enough to stimulate a response and affect the issue at stake.

According to de Jong, "radical political activist movements" are becoming increasingly prolific through the use of new communications technology (de Jong et al., 2005).

Activist politics have thus significantly influenced mainstream politics over issues such as the environment (de Jong et al., 2005).

The first Norwegian environmental NGOs were, in the main, "traditional" organisations largely financed by their members, who also elected the leaders. Since the late 1980s, though, a different type of organisation has emerged. Organisations such as Bellona and Norges Miljøvernforbund were (and still are) dominated by strong leaders, and focus on activism to attract media attention. Their funding comes principally from sponsors which are often businesses wanting to portray themselves as environmentally responsible (Arntsen, 2004; Eidsmo, 2003; Hansen, 2007; Johnsen, 1996a, 2006; Jørgensen, 2000; Lange et al., 2002; Løken, 1999; Løne, 2001; Nilsen, 1997; Notøy, 2002; Østrem, 1988; Roness, 1996; Rødland and Oddekalv, 2006; Røkeberg, 1996; Sæten, 1991; Selle, 1991; Søgård, 1995; Statle, 1993; Strømsnes, 2001, 2003; Wiker, 1990).

On the international stage, Greenpeace, for example, became very successful in attracting media attention and establishing itself as a transnational environmental organisation (Brown and May, 1989; Lucas and Dorrian, 2006; Nordskag, 1999). Greenpeace was also successful in adapting to new media and new marketing possibilities (Hart et al., 2005) and it has received acclaim for the daily updates on its website (de Jong et al., 2005, p. 115). Greenpeace and other transnational NGOs have historically worked closely with journalists (Weyler, 2004) and there are many examples of their growing influence and power in the 1970s, 1980s and 1990s (Gidley and Shears, 1986; Morgan and Whitaker, 1986).

Sampei and Aoyagi-Usui (2009) have found a correlation between increased newspaper coverage and public concern for global warming. Growing concern, however, does not necessarily lead to action. According to Hulme, sensationalist and frightening newspaper headlines can be counter-productive (Hulme, 2007), and he rebukes the British media for exaggerating the risks in their reporting on climate change. Such reporting is seen to contribute to "a climate of fear about our future climate" (Hulme, 2008) and could possibly lead to denial or apathy, according to Dilling (Dilling and Moser, 2007). According to Boykoff (2008), "contextualisation within climate-science stories helps to ward off misdirected controversy and to sort marginalised views from counter-claims worthy of consideration".

In the literature on media and climate change, however, the employment of norms of objectivity and balanced reporting have received much criticism. According to Boykoff and Boykoff, in the United States the norm of "balanced reporting" has had a rather detrimental effect on the media coverage of climate science and other scientific topics (Boykoff and Boykoff, 2004), in part because balance is often a substitute for validity checks. According to Dunwoody and Peters, the typical journalist has "neither the time nor the expertise to check the validity of claims" (1992, p. 210). Balance without validity checks can lead to unbalanced reporting, according to Mooney (2004), giving undue attention to marginal views held by very few of the experts. Balance thus sometimes comes in conflict with the ethics of journalism such as, for instance, the responsibility for giving the public the information necessary for their protection against dangers. Boykoff claims that the way journalists employ norms related to personalisation, dramatisation, and novelty, "significantly influence" the employment of other types of norms—for instance, the norm of "balanced reporting" (Boykoff and Boykoff, 2007). On the other hand, different studies have indicated that media elsewhere do not necessarily employ journalistic norms in the same way as do the US media (Boykoff and Rajan, 2007).

Methodology

The study takes an empirically driven and exploratory approach where quantitative findings are employed as a basis for reflection on possible causes. The first part of this study is a quantitative investigation based on retriever.no, a searchable database containing the leading Norwegian newspapers. The database makes it possible to search for keywords in articles, thus tracking developments over time. The first part of the study follows 17 Norwegian NGOs from January 1999 to January 2009, and compares how often they were quoted in the media.[1] A period of 10 years makes it possible to run regressions in order to compare trends in visibility over time. The number of articles quoting or mentioning the organisation is counted, but further division and categorisation does not take place at this stage.

A broad range of organisations has been selected for the study sample. In order to test the results for other explanations, care has been taken to also include organisations or institutions not strictly, or not exclusively, within the field of environmental organisations. The selection includes relatively "old" and newer NGOs. Some of the NGOs focus exclusively on the environment, but most are also engaged in development assistance in the South. Some are mainly engaged in research, while others are more geared towards public relations and activism. The Center for International Climate and Environmental Research—Oslo (CICERO) is also included, although CICERO is not an NGO. It was established by the Norwegian government and is now an independent research centre associated with the University of Oslo. Nonetheless, including a research-based institution like CICERO gives depth to the investigation and facilitates analysis of the significance of science and scientists in the media. This will help in the analysis of possible causal explanations for the results found for the NGOs. Organisations such as Save the Children have also been included to test for causal mechanisms related to trends in media coverage of development assistance in general.

The selected organisations are: Utviklingsfondet (The Development Fund), Kirkens Nødhjelp (Norwegian Church Aid), Redd Barna (Save the Children), Norsk Folkehjelp (Norwegian People's Aid), Framtiden i våre hender (The Future in Our Hands), Regnskogfondet (Rainforest Foundation Norway), CICERO, Zero, Greenpeace, Norges Røde Kors (Norwegian Red Cross), Operasjon Dagsverk (Operation Dayswork), Norges Naturvernsforbund (Friends of the Earth Norway), Norges Miljøvernforbund (Green Warriors of Norway), Bellona, Changemaker, World Wide Fund for Nature (WWF) and Natur og Ungdom (Nature and Youth).

A few words of caution are necessary regarding the reliability of this part of the investigation. First, the number of newspapers in the database has increased significantly over the 10 years and, in addition, some newspapers have reduced the number of pages in print editions or changed from broadsheet to tabloid format, which significantly affects the total number of articles and the total number of words in each edition. Meanwhile, most newspapers have significantly expanded online editions over the 10 years. It is therefore reasonable to expect variances, over time, in the number of times a name or a theme is mentioned in the newspapers.

To estimate the effects of these and other factors, a test has been made. The occurrences of common Norwegian words such as "er" ("is" or "are") and "år" ("year") has been counted and the use of these common words was found to have increased by between 100 and 150 per cent between 1999 and 2009. Testing for typical words in the vocabulary of the selected organisations—for example "climate", "climate change", "development assistance" and "poverty"—we find the use of these words increasing two to three times faster than the common words ("er", "år"), indicating a generally increasing interest in the environment, in climate and in development assistance. Development assistance has been included for two reasons: first, because many of the NGOs studied here also are engaged in development assistance or poverty reduction in the South (these two themes are also often connected in the discourses the NGOs employ) and, second, these two words are often also connected in the general, public discourses on the environment and climate change in Norway. The tests show that it is reasonable to expect a significant growth in the number of times an organisation is mentioned in the newspapers, most likely somewhere between 200 and 300 per cent. These introductory tests also indicate that the investigation contains a relatively large margin of error. The results should therefore be used only to describe some very general trends.

More investigation is needed before an analogy from these results may be made to the media in Norway. The results from the quantitative part are systematised using Qualitative Comparative Analyses (QCA)—for more on that see below—to sort out the combination of variables that can be said to have determined success or failure in setting the agenda on climate change in Norway.

Two more tasks have been developed using qualitative methodology in order to contextualise the findings of the quantitative analysis. In the first place, all newspaper articles concerning the climate change conference in Poznan in December 2008 are investigated. The articles are categorised according to several criteria: the type of article, the name of the journalist, and the type and names of the sources. A closer qualitative reading is made in order to understand the argument expressed, explicitly or implicitly, in the articles; the purpose is to provide new knowledge about trends in the coverage of climate change issues in the Norwegian media. The development of the second task is aimed at a better understanding of the media strategy of one of the organisations that seems to be successful in setting the agenda. The Development Fund is investigated, employing interviews with key informants in the organisation. This period includes activities at Poznan and the launch of a major report.

Quantitative Investigation

The first overview of the results shows substantial variation between the different NGOs. Some have, apparently, been much more successful in increasing their visibility in the media than others. Some organisations have seen a significant increase in the number of articles where the organisation is mentioned, sometimes by as much as 1000 per cent over the period. Others, by contrast, have seen the number of such mentions stagnating.

The organisations are divided into three categories for analysis: "losers", "winners" and "average performers", based on the results presented in Table 1. The "losers" show little or no increase over the 10 years (increases of less than 100 per cent are defined as a relatively poor result). The "winners", by contrast, show growths of between 200 and 1000 per cent.

Five organisations have been categorised as losers, among them some major Norwegian NGOs with large membership bases and long traditions—for example, the Norwegian Red Cross and Norwegian People's Aid. The Norwegian Red Cross is mentioned approximately the same number of times in 2008 as it was 10 years earlier, and what this means in reality is that the Norwegian Red Cross is significantly less visible today than it

TABLE 1
Number of articles mentioning a topic or an organisation

Topic/name of organisation	1999	2000	2001	2002	2003	2004	2005	2006	2007	2008
Is ("er")	238,485	278,688	301,077	308,558	372,438	379,610	458,775	508,677	528,226	575,547
Year ("år")	119,564	135,610	142,891	148,828	175,064	179,382	213,769	232,410	242,962	270,738
Global warming	69	154	186	140	112	177	351	596	1609	1029
Climate change	25	77	59	68	37	70	90	153	310	208
The greenhouse effect	28	70	75	46	34	49	94	104	120	80
Development aid	1420	1478	1947	2048	2102	2171	3036	3197	3109	3774
Politics	7588	7818	9782	10,083	11,770	10,881	14,276	14,868	16,178	17,471
Bellona	390	549	531	510	379	449	559	815	1036	1027
Changemaker	7	10	29	51	73	84	93	80	108	139
CICERO	59	59	81	60	66	93	139	160	495	476
Friends of the Earth Norway	259	229	373	210	279	316	332	500	561	502
Greenpeace	237	268	430	302	291	179	258	295	361	306
Nature and Youth	208	394	452	478	510	449	605	861	741	867
Green Warriors of Norway	187	145	131	131	94	117	102	162	209	203
Norwegian Church Aid	300	254	442	287	489	620	711	543	607	816
Norwegian People's Aid	503	380	483	402	455	495	571	560	489	744
Norwegian Red Cross	479	268	409	434	592	783	856	453	410	395
Operation Dayswork	44	49	69	72	86	100	89	100	118	151
Rainforest Foundation Norway	7	5	24	27	26	29	74	45	146	127
Save the Children	483	326	444	397	792	718	943	587	614	836
The Development Fund	17	12	28	38	69	94	154	132	127	215
The Future in Our Hands	117	119	202	220	244	219	304	327	457	432
World Wide Fund for Nature	126	126	282	271	303	294	414	507	686	546
Zero	77	88	169	312	245	299	403	527	531	537

was 10 years ago. It is engaged in voluntary work in Norway and in development assistance, but especially in emergency assistance, while Norwegian People's Aid is probably best known for development assistance and solidarity with the South, including sustainable development and activities to protect the environment. Norwegian People's Aid shows some increase in press citations, but much less than most of the other organisations. Save the Children performs slightly better than these first two, but still does relatively poorly compared with other organisations (Save the Children is mainly engaged in development assistance in the South). These three organisations are all large, well funded and well known.

The last two losers are relative newcomers and, in many ways, very different from the first three. Green Warriors of Norway is the most activism-driven organisation in the selection; it is relatively young, has few members and has one dominating leader. Although the Green Warriors of Norway is responsible for some of the most visible environmental activism in Norway over the last 10 years, this does not seem to have helped them to maintain a sustained media interest. It is a small organisation and it is only to be expected that attention from the media fluctuates. The long-term trend, though, is of only a very small increase, approximately 25 per cent, much less than might be expected. Greenpeace is the last of the losers. In the light of the existing literature presented above, it is surprising to find Greenpeace in this category. A closer look at the results for Greenpeace shows that it performed particularly poorly after 2001, falling more than 20 per cent over the last seven years of the study. Greenpeace is a relatively young NGO in Norway, established in 1988, and currently employs five in the Norwegian office. Despite its focus on non-violent activism and the dissemination of information (Greenpeace, 2009), Greenpeace seems to have become less and less relevant for Norwegian newspapers.

These results become even more intriguing when contrasted with the group of "winners". Among the winners we find eight very different organisations. Six can be defined as membership-driven, while two are mainly dedicated to research and the dissemination of information. The first six winners are The Development Fund, Changemaker, WWF, The Future in Our Hands, Rainforest Foundation Norway, and Nature and Youth—all, paradoxically, membership-run organisations that have struggled to attract new members in the same period—as have most Norwegian organisations, with falling membership numbers over the last 10–15 years. This does not, however, signify that these organisations are becoming less relevant as sources for journalists. The majority of the NGOs combine an interest in environmental issues with an interest in fighting poverty in the South. According to existing literature, all these organisations pay substantial attention to internal procedures and democratic participation. Some, Nature and Youth, and Changemaker in particular, combine internal education, democratic participation and activism.

The last two winners, CICERO and Zero, are, again, very different from the other organisations in that they focus almost exclusively on research and on disseminating information. While they do not engage in activism as such, they nonetheless participate actively in the public debate, and their researchers regularly call for action to protect the environment. They have been included to test media interest more broadly, for researchers whose agenda is the environment in general and climate change in particular. The results show that such sources are quoted more and more frequently in Norwegian newspapers as the media have increasingly paid more attention to climate change issues.

More investigation is needed into each of these organisations and their media strategies in order to draw definitive conclusions about the reasons for success and failure in attracting media attention, but it is nonetheless possible to construct some hypotheses based on the patterns found in this material.

In the first place, a large organisation staffed with people competent in information management and public relations is not necessarily successful in attracting media interest. This could possibly be due to the increasing attention media organisations and journalists pay to the role and problems connected with information management and public relations which they see from the perspective of journalistic norms on independence.

Secondly, those organisations that pay the most attention to activism and media friendliness, Greenpeace and Green Warriors of Norway, may have made themselves less attractive as sources for the media over the 10 years of the study, and this is not what might be expected from the literature presented earlier. According to the literature on Norwegian media and the environment, in-depth journalism on the long-term effects of environmental issues was generally becoming less likely because the media ignore environmental issues or focus only on spectacular news (Johnsen, 1996b; Mørck, 1996). The material presented here does not refute such an argument, but it does indicate that other, and possibly more complex, processes are in play.

Thirdly, membership-run youth organisations such as Changemaker, and Nature and Youth show consistent growth over the 10 years. More study is required before a conclusion is made about the causes, but one hypothesis for future investigation is related to the recruitment of journalists. Anecdotal evidence indicates that many journalists and editors are former members of these youth organisations, so it is possible that they are more likely to use members of the same organisations as sources when they are covering environmental issues.

A fourth hypothesis is that six of the winners have relatively small staffs, and have thus chosen to focus on a limited number of issues, seeking niches where they have become particularly useful as sources for journalists. At the same time, these sources provide the journalists with clear-cut standpoints, often in contrast to those of researchers, and in opposition to those of politicians.

Yet another possible reason for the failure to attract media attention is that climate change has emerged as a dominant environmental issue. In contrast to pollution or other local concerns, climate change is primarily a global issue that is not easily captured using the activist media strategies employed by some of the organisations in the 1980s and 1990s. This notion will be investigated further below.

Finally, with increasing attention being paid to climate change issues, the debate on the environment has become more complex, involving many factors, many specialised disciplines and knowledge—and very few people have sufficient knowledge of the topic to make sense of the new information on climate change.

Qualitative Comparative Analysis to Systematise the Results

From the above, it is apparent that at least four different variables must be considered in the analysis of possible causal mechanisms explaining success or failure in setting the agenda in the Norwegian media. The first is that the type of activities the NGO undertakes seems to have an effect, and among the NGOs are some that specialise in the spectacular to attract media attention, and some that do not. The second is that there are a number of traditional organisations with strong internal institutions and procedures for democratic participation, and some that belong to a different category. Third, it seems to be important whether the NGO focuses exclusively on climate change; for some NGOs, climate change is only one of many concerns. And finally, whether or not the NGO is also engaged in development assistance to the South also seems significant.

Qualitative Comparative Analysis (QCA) is a comparative tool that employs Boolean algebra "for a systematic analysis of similarities and differences across cases" (Rantala and Hellström, 2001, p. 87). Employing Boolean algebra to comparative studies has become a much used methodology in social sciences over the last few years (Byrne and Ragin, 2009; Ragin, 1987, 2000, 2008; Ragin and Amoroso, 2011; Rihoux and Ragin, 2009).[2] Comparison is a potentially useful tool in the investigation of causal mechanisms for journalism studies also. But according to Ragin (1987) it is almost impossible to compare more than between four and eight different cases, when each case consists of four or five variables, using traditional qualitative methodologies, because the number of possible combinations of variables becomes too big. Therefore, following Ragin (1987), a holistic Boolean approach will be taken. Since this a relatively new methodology in journalism studies, a brief account of the procedure will be given in the following section, including the variables that were applied and how they were operationalised.

Based on the statistical analysis, four possible causes for explaining increasing or stagnating visibility in the newspapers have been formulated:

1. A possible cause might be related to the media strategy of the various organisations. Some have employed a strategy focusing on media-friendly spectacles, while others have been more geared towards other types of activities.
2. Another possible cause might be found in the type of organisation. There are at least two types of organisations in the selection of organisations.
3. There also seems to be good reason to hypothesise a causal relationship between focusing on global climate change and increased visibility in the newspapers.
4. The quantitative analysis also seems to indicate a possible causal relationship between engagement in development assistance and increased visibility.

The advantage of QCA employing Boolean algebra is (among other things) that it does not see these possible causal explanations as mutually excluding. Instead, it facilitates a holistic analysis that opens up for the possibility that more than one cause must be accounted for in order to explain a specific outcome for one or several organisations.

The first step of Boolean algebra is to simplify. Each NGO will be seen as a case. Each case will be ascribed a value for each of the chosen variables. The value here can only be 0 (not present) or 1 (present). An NGO making use of spectacle to attract media attention will be ascribed the value 1 to the variable "spectacular". This is of course a very raw simplification. In reality, most NGOs would be somewhere between 0 and 1 on a sliding scale. But going through the process of simplification nonetheless has some important benefits because it helps the researcher to focus systematically on the differences between the cases, highlighting possible explanations for the findings. Once completed, the results can be further analysed using Boolean algebra, reaching the most accurate and simplest valid statement of causal relationships between variables based on the empirical evidence and the simplifications. Finally, the process makes the reasoning transparent. It becomes possible for others to test the findings. To make the coding of this study more transparent, I include a short description of how the different NGOs have been coded in relation to the chosen variables.

Five organisations have been categorised as using spectacle to attract media attention, and are thus given the value 1 for the variable "spectacular": Green Warriors of Norway, Greenpeace, Bellona, Nature and Youth, and Changemaker. The rest are given the value 0 for this variable.

Five organisations have been categorised as new, and less focused on traditional organisational values as regards democratic participation. They are given the value 0 for the variable "organisation": CICERO, Zero, Green Warriors of Norway, Greenpeace and Bellona. They stand in contrast to the rest that are deemed to be more attentive to procedures and institutions for democratic participation (thus given the value 1).

The majority of the organisations can be said to be increasingly concerned with global climate change, and are thus given the value 1 for the variable "climate change". Seven of the organisations are less noticeable in the climate change debate and have been given the value 0 for this variable: Norwegian Church Aid, Save the Children, Norwegian People's Aid, The Future in Our Hands, Green Warriors of Norway, Norwegian Red Cross, and Operation Dayswork. The rest are given the value 1 for this variable.

Eight of NGOs are deemed to be considerably more engaged in development assistance to the South than the other nine and they are given the value 1 for the variable "development assistance": The Development Fund, Norwegian Church Aid, Save the Children, Norwegian People's Aid, Rainforest Foundation Norway, Norwegian Red Cross, Operation Dayswork, and Changemaker. The others are deemed to be most visibly engaged in activities in Norway and are thus given the value 0.

The next step is to plot these cases, the NGOs, with their respective combinations of factors, in a so-called "truth table". This has been achieved using the Karnaugh Minimiser, first for the NGOs deemed to have been successful in setting the agenda, and thereafter for those which have struggled, resulting in two sets of Boolean expressions. The first is a typology of the successful NGOs while the other is a typology of the struggling NGOs.

In this material are three types of NGOs that have been evaluated as successful in setting the agenda on the environment in Norway.

The first type is defined by the following combination of variables: they do not employ spectacle to attract media attention. They focus on climate change but not on development assistance. For this type, dedication to procedures and institutions for democratic participation does not seem to matter; their success is independent of this particular variable.

The second type is different. It is defined by its dedication to procedures and institutions for democratic participation, which can be said to be important for success in setting the agenda. But this type is not particularly engaged in climate change. And in contrast to the former type, it is engaged in development assistance.

The third type is again defined by its dedication to procedures and institutions for democratic participation. This type does not employ spectacular activities to attract media attention, and is not engaged in development assistance. Its success can be said to be independent of any engagement in climate change issues.

The results from the Boolean analysis of the NGOs struggling to get media attention also define three types of organisations based on the combinations of variables.

The first type of NGO struggling for media attention is defined by the following combination: it employs spectacle to attract media attention and it focuses on climate change issues, but not on development assistance. The next type of struggling NGO is defined by three variables. It also employs spectacle to attract media attention. It is not particularly dedicated to procedures and institutions for democratic participation within the organisation, and is not engaged in development assistance. It struggles to attract media attention whether or not it focuses on climate change. The third type of struggling NGO does not employ spectacle to attract media attention, is dedicated to internal democratic participation and focuses on development assistance, but not on climate change.

With the use of Boolean algebra, 17 different cases have been logically reduced to six generic types, each defined by a specific combination of variables seen as causing success—or failing to prompt success—in setting the agenda in Norwegian media, leading to the conclusion that some variables seem to have been especially influential in deciding the outcomes. Spectacle to attract media attention appears in many unsuccessful types of NGOs whereas a strong dedication to internal democratic participation is found in many successful types. A focus on climate change seems to make a positive impact on media attention. A focus on development assistance also seems to make an impact, but only when combined with other, more influential, variables. The Boolean algebra thus verifies, to some extent, the hypothesis presented in the previous section. But it also highlights the importance of taking the contexts into account.

While the investigation has thus far yielded interesting results about the Norwegian NGOs and their relationship to the media, there remain many unanswered questions. The quantitative studies and simplification of Boolean algebra can only indicate general tendencies, and to learn more about why different NGOs are successful or not successful it is necessary to conduct qualitative studies of individual relationships with the media. More research on the individual NGOs will produce more knowledge on, for instance, why dedication to procedures and institutions for democratic participation within the organisation seems to be so important. The following is meant as a small beginning of such an endeavour in that they focus on some of the more intriguing questions raised by this investigation so far: Why do procedures and institutions for democratic participation within the organisation seem to matter? What role does climate change play? I try to sketch a possible answer to these questions by looking into one NGO that stands out as successful.

The Newspaper Coverage of Poznan

Some possible explanations for the findings have been discussed above. The emergence of climate change as a dominant issue in the media coverage of the environment was presented as one of the hypotheses. The use of the term "global warming" ("*global oppvarming*") increased by approximately 700 per cent between 2000 and 2007, before falling slightly after the global conference on climate change in Bali. The terms "greenhouse effect" and "climate change" show a similar development. These results clearly show an increasing interest in and coverage of climate change issues in general.

The 14th Conference of the Parties to the United Nations Framework Convention on Climate Change (UNFCCC) and the 4th Session of the Meeting of the Parties to the Kyoto Protocol took place in Poznan in December 2008. To investigate some relevant trends in Norwegian media coverage of climate change, all articles on the conference in Poznan registered on retriever.no were studied and categorised according to type of article, main sources, arguments on climate change issues and type of meta-narrative (if any) in the article. The research shows significant differences between the newspapers and the analysis therefore needs to start with a short description of the Norwegian newspapers covering the Poznan conference.

The conference at Poznan was less dramatic than its predecessor at Bali in 2007, which led to much less media coverage of it, especially in the large national newspapers. Most articles appeared in small national papers or larger regional newspapers, *Adresseavisen* being the newspaper publishing most articles, followed by *Klassekampen*, *Nationen* and *Vårt Land*.

Being a regional newspaper, *Adresseavisen* uses its identity as a member of the local community ("*trønder*") as a news criterion, and published 25 per cent of the articles on the conference in Poznan principally to tell the community of "*trøndere*" about other "*trøndere*" participating in the conference. Nonetheless, 75 per cent of the articles are grounded in an argument of "importance"—the newspaper is telling its readers to read this because it is important for them. Importance is combined with a meta-narrative well known from other studies of Norwegian journalism on international issues: the story of Norway's altruistic struggle for the good of all (Eide and Simonsen, 2008; Krøvel, 2011; Tvedt, 1998, 2002, 2003). *Klassekampen*, meanwhile, covered the conference differently. Here we find no patriotism, but much criticism of Norway and the role it played. All the articles are submerged in the same meta-narrative: climate change is real, caused by human activity, and is a matter of survival for all humankind. Again, "importance" is the dominating news criterion of journalists and editors. The principal sources are academics and other experts on climate change, or else activists particularly known for their knowledge of the subject. Other newspapers, like *Stavanger Aftenblad*, *Dagbladet* and *Ny Tid*, focus mainly on producing commentaries from their home ground, often inviting scientists or activists to publish in-depth observations. Again, two types of sources dominate: experts and "activist-experts".

The investigation of the Norwegian media coverage of the conference in Poznan shows a pattern: the articles provide evidence of knowledge of climate change issues, indicating that journalists have had enough time to investigate and evaluate sources and contrasting claims before producing articles. According to Dunwoody and Peters (1992), journalists normally have neither the time nor the expertise to uphold the ideal of assessing the validity of claims or arguments, but journalists covering Poznan must have spent a lot of time investigating before reporting. Then, many of the expert sources were sought out long before the conference, and many journalists did not even participate (being at the conference was not a necessary condition for being quoted in articles on the conference). A third point is that most of the journalists took a stand before covering the conference, seeking, in the words of Goldman (1999), to uphold a "veritistic" ideal, trying to evaluate conflicting claims and avoiding using the norm of balance as a substitute for validity checks, as has been found elsewhere (Boykoff, 2007; Boykoff and Boykoff, 2004). This means that only a limited number of sources are found in the media coverage. The majority of sources are quoted because of their expertise, and the NGOs mentioned or quoted are, in general, only those which are particularly knowledgeable on climate change. Hardly any sources are quoted because of activism, demonstrations, protests or other activities to attract media attention.

However, this does not mean that "identification" has disappeared as a news criterion. The regional newspapers, in particular, employ "identification" to frame the information when selecting sources and explaining the issue to the readers. But compared with the results of other studies, first-order journalistic norms, for instance "identification" and "dramatisation", play a relatively small role in comparison with second-order norms such as checking whether the story holds (Boykoff and Boykoff, 2007).

Results Pertaining to The Development Fund

The analysis above might explain the success found for some of the NGOs, for instance Zero, CICERO and—to a lesser extent—Friends of the Earth Norway and Bellona. For several years, these organisations have specialised in climate change-related topics. Others in the group of "winners", though, are harder to explain from a climate change perspective. The Development Fund, for instance, combines concern for the environment with solidarity-based development assistance, but is not particularly known for its expertise on climate change. A study of the media strategy of The Development Fund is therefore useful for providing better understanding of the relationship between the media and environmental NGOs.

The Development Fund dedicated substantial resources to the Poznan conference, sending two experienced members of staff and producing numerous newsletters and reports from the conference on relevant topics (Sigurd Jorde, interview, 10 October 2008; Aksel Nærstad, interview, 5 January 2009). But The Development Fund did relatively poorly in setting the agenda in the Norwegian media as regards the Poznan conference. In the articles about the conference in Poznan it was only quoted once—and this strengthens the hypothesis that the journalists covering climate change tend to seek out those sources deemed particularly knowledgeable on the issue. The Development Fund found it very difficult to move from being perceived a valuable source on many other environmental issues to being accepted as a useful source on climate change in particular. The position as "activist expert" was monopolised by a handful of NGO representatives who had already established relationships with journalists on this particular issue.

The next point is that the long-term results on The Development Fund and the media coverage over the 10 years studied here show that two or three Development Fund representatives dominate, and are the sources of the vast majority of the articles in which The Development Fund is mentioned or quoted. A closer analysis reveals that these representatives each have one or two areas of expertise and in most of the articles they provide the journalists with alternative perspectives to those given by Norwegian officials. Unlike most scientists, they formulate clear policy positions and propose actions to resolve or reduce a problem.

The third point is that The Development Fund launched a number of other activities that can help to explain the long-term success in setting the agenda. The youth organisation Spire was engaged in internal educational activities, attracting many new members and constantly increasing their level of knowledge on environmental issues. Spire put together workshops on writing articles and other topics relevant for the media strategy. At the same time, The Development Fund had in place an advanced system for media surveillance and routines for analysing the results. The organisation paid a great deal of attention to learning and improving from experience.

Additionally, considerable effort was made to produce reports on issues particularly important for the organisation. In March 2009, The Development Fund launched the report *En verden uten sult* (*A World Without Hunger*) on the environment, development assistance and global politics. The organisation used the report on several levels: to educate the members themselves; to attract new members (more than 200 people showed up at a seminar when the report was launched); and to attract media attention (although it resulted in only a limited number of news articles).

The effect of the above was not immediate media coverage of demonstrations, activism or protests. But the long-term effect was that The Development Fund succeeded in identifying some members as knowledgeable and useful sources for the media. Another long-term effect was a constant flux of members and former members educated in the handling of media and in environmental issues. Some members subsequently became journalists themselves, covering and commenting on environmental issues. More investigation is needed to shed light on the long-term effect of this process.

Conclusion

The results presented in this article are enlightening when seen in relation to earlier studies on the media and environmental issues. These results do not support earlier findings that highlight increasing media focus on sensation and identification, and criticise the coverage for lack of depth and quality (Johnsen, 1994, 1996b; Mørck, 1996). By contrast, the "winners" found in this exploration are in general those organisations focusing on producing knowledge, and not those focusing on supposedly media-friendly activism.

These results also stand in contrast to existing literature on the success of Greenpeace (Brown and May, 1989; Gidley and Shears, 1986; Weyler, 2004). Greenpeace and similar organisations found it hard to compete for media attention as the media interest turned from pollution, roads, dams and local matters to the very complex question of global climate change.

The results also seem to contrast with earlier findings on balance as a substitute for validity checks. The newspapers covering Poznan dedicated a few selected journalists to research climate change in more detail, conforming with Goldman's (1999) "veritistic ideal".

According to Harcup (2009), objectivity (checking whether a story holds) can be seen as a ritual in journalism. The results of this investigation indicate a normative shift—away from merely presenting conflicting claims to attempting to evaluate conflicting claims, as also indicated in other recent studies of Norwegian journalism on climate change (Ytterstad, 2008). Conflict, identification and dramatisation seem to play a less important role in Norwegian newspaper journalism about climate change than might be expected as a result of previous studies.

NOTES

1. Only newspapers registered on retriever.no by January 2009.

2. For more detailed information on the QCA and Boolean algebra, I refer the reader to Byrne, Hicks and Ragin (Byrne and Ragin, 2009; Hicks, 1994; Ragin, 1987; Ragin and Amoroso, 2011; Rihoux and Ragin, 2009).

REFERENCES

ARNTSEN, BJØRNULF (2004) *Konflikt og konsensus: en studie av mål, virkemidler og organisering innenfor Natur og Ungdom og Bellona*, Oslo: Oslo University.

BOYKOFF, MAXWELL T. (2007) "Flogging a Dead Norm? Media Coverage of Anthropogenic Climate Change in United States and United Kingdom, 2003–2006", *Area* 4(39), pp. 470–81.

BOYKOFF, MAXWELL T. (2008) "The Real Swindle", *Nature Reports Climate Change*, 21 February, http://www.nature.com/climate/2008/0803/full/climate.2008.14.html, accessed 1 February 2011.

BOYKOFF, MAXWELL T. and BOYKOFF, JULES M. (2004) "Balance as Bias: global warming and the US prestige press", *Global Environmental Change* 15(2), pp. 125–36.

BOYKOFF, MAXWELL T. and BOYKOFF, JULES M. (2007) "Climate Change and Journalistic Norms: a case-study of US mass-media coverage", *Geoforum* 38(6), pp. 1190–204.

BOYKOFF, MAXWELL T. and RAJAN, S. RAVI (2007) "Signals and Noise: mass media coverage of climate change in the USA and the UK", *EMBO Reports* 8(3), pp. 207–11.

BROWN, MICHAEL and MAY, JOHN (1989) *The Greenpeace Story: "the optimism of the action is better than the pessimism of the thought" Harald Zindler*, London: Dorling Kindersley.

BROWN, MICHAEL HAROLD and MAY, JOHN (1991) *The Greenpeace Story*, New York: Kindersley.

BYRNE, DAVID S. and RAGIN, CHARLES C. (2009) *The Sage Handbook of Case-based Methods*, Los Angeles and London: Sage.

DE JONG, WILMA, SHAW, MARTIN and STAMMERS, NEIL (2005) *Global Activism, Global Media*, London: Pluto.

DILLING, LISA and MOSER, SUSANNE C. (2007) *Creating a Climate for Change: communicating climate change and facilitating social change*, Cambridge: Cambridge University Press.

DUNWOODY, SHARON and PETERS, HANS PETER (1992) "Mass Media Coverage of Technological and Environmental Risks: a survey of research in the United States and Germany", *Public Understandings of Science* 1(2), pp. 199–230.

EIDE, ELISABETH and SIMONSEN, ANNE HEGE (2008) *Verden skapes hjemmefra: pressedekningen av den ikke-vestlige verden 1902–2002*, Oslo: Unipub.

EIDE, ELISABETH and YTTERSTAD, ANDREAS (2009) "The Tainted Hero—Norwegian representation of the Bali conference", paper presented at the Global Dialogue Conference, Århus.

EIDSMO, BJØRNAR (2003) *Samfunnsansvarlig partnerskap: synergieffekter og veier til suksess*, Oslo: Oslo University.

GIDLEY, ISOBELLE and SHEARS, RICHARD (1986) *The Rainbow Warrior Affair*, London: Unwin Paperbacks.

GOLDMAN, ALVIN I. (1999) *Knowledge in a Social World*, Oxford: Clarendon Press.

GREENPEACE (2009) "Om oss", http://www.greenpeace.org/norway/about, accessed 10 October 2009.

HANSEN, ANE BJØLGERUD (2007) *Oppbrudd fra forbrukssamfunnet: bevegelsen Framtiden i våre hender 1974–1982*, Oslo: Oslo University.

HARCUP, TONY (2009) *Journalism: principles and practice*, 2nd edn, London: Sage.

HART, TED, GREENFIELD, JAMES M. and JOHNSTON, MICHAEL W. (2005) *Nonprofit Internet Strategies: best practices for marketing, communications, and fundraising success*, Hoboken, NJ: Wiley.

HICKS, ALEXANDER (1994) "Qualitative Comparative Analysis and Analytical Induction", *Sociological Methods & Research* 23(1), pp. 86–113.

HULME, MIKE (2007) "Newspaper Scare Headlines Can Be Counter-productive", *Nature* 445(7130), p. 818.

HULME, MIKE (2008) "The Conquering of Climate: discourses of fear and their dissolution", *Geographical Journal* 174(1), pp. 5–16.

IHLEN, ØYVIND (1999) *Medier og miljø: en skisse av norske avisers miljødekning 1977–1997*, Oslo: Universitetet i Oslo, Institutt for medier og kommunikasjon.

IHLEN, ØYVIND (2001a) "Medier, miljø og påvirkning" ["Media, Environmental Issues, and Effects"], *Tidsskrift for samfunnsforskning* 42(1), pp. 65–88.

IHLEN, ØYVIND (2001b) "Miljømakt og journalistikk: Retorikk og regi" ["Environmental Power and Journalism: rhetoric and direction"], in: Martin Eide (Ed.), *Til dagsorden! Journalistikk, makt og demokrati [To the Agenda! Journalism, power and democracy]*, Oslo: Gyldendal Akademisk, pp. 304–28.

JOHNSEN, JAN (1994) *Massemedier og miljø: kompendium: natur, samfunn, miljø*, Oslo: Høgskolen i Oslo, Avdeling for lærerutdanning.

JOHNSEN, JAN (1996a) *Miljø, medier og skole*, Oslo: ProSus.

JOHNSEN, JAN (Ed.) (1996b) *Rapport nr. 3/96: Miljø, medier og skole*, Oslo: Prosus, Norges Forskningsråd.

JOHNSEN, JAN (2006) *Miljø, medier og skole: rapport nr. 3/96*, Oslo: Pensumtjeneste.

JØRGENSEN, JAN (2000) *Den miljøpolitiske definisjonskampen: en case-studie av miljøorganisasjonen Framtiden i våre hender*, Bergen: University of Bergen.

KRØVEL, ROY (2011) "Journalistic Narratives of Success and Failure at the Bali Climate Change Conference in 2007", *Intercultural Communication Studies* 20(2), pp. 89–105.

KWANSAH-AIDOO, KWAMENA (2001) "The Appeal of Qualitative Methods to Traditional Agenda-setting Research: an example from West Africa", *International Communication Gazette* 63(6), pp. 521–37.

LANGE, SIRI, SPISSØY, ARILD and BRUDVIK, MARIE (2002) *Fra motstander til medspiller: partnerskap mellom norske frivillige organisasjoner og norsk næringsliv*, Bergen: Chr. Michelsen Institute.

LØKEN, ANNE SOFIE (1999) *Veier til et bærekraftig forbruk: en komparativ analyse av virkelighetsoppfatning og virksomhet i Norges naturvernforbund og Framtiden i våre hender*, Oslo: Universitetet.

LØNE, CECILIE (2001) *Communicating Solutions for a Greener World*, Oslo: Oslo University.

LUCAS, GAVIN and DORRIAN, MICHAEL (2006) *Guerrilla Advertising: unconventional brand communication*, London: Laurence King.

MOONEY, CHRIS (2004) "Blinded by Science: how 'balanced' coverage lets the scientific fringe hijack reality", *Columbia Journalism Review* 43(4), pp. 26–35.

MØRCK, SIDSEL (1996) "Tilslører eller avslører mediene grunnleggende miljøproblemer i vår tid?", in: Jan Johnsen (Ed.), *Miljø medier og skole. Rapport nr 3*, Oslo: Prosus, pp. 15–9.

MORGAN, ROBIN and WHITAKER, BRIAN (1986) *Rainbow Warrior: the French attempt to sink Greenpeace*, London: Hutchinson.

NILSEN, ANN (1997) *Miljøsosiologi: samfunn, miljø og natur*, Oslo: Pax.

NORDSKAG, MORTEN (1999) *Grønn og global: en analyse av Greenpeace som transnasjonal aktør i den internasjonale handelen med farlig avfall*, Oslo: Universitetet i Oslo.

NOTØY, GRETE (2002) *Bellona: fra nasjonal aktivist til internasjonal aktør*, Molde: Høgskolen i Molde.

OSTREM, HANS CHR. (1988) *Miljøstiftelsen Bellona*, Oslo: Norsk journalisthøgskole.

RAGIN, CHARLES C. (1987) *The Comparative Method: moving beyond qualitative and quantitative strategies*, Berkeley: University of California Press.

RAGIN, CHARLES C. (2000) *Fuzzy-set Social Science*, Chicago: University of Chicago Press.

RAGIN, CHARLES C. (2008) *Redesigning Social Inquiry: fuzzy sets and beyond*, Chicago: University of Chicago Press.

RAGIN, CHARLES C. and AMOROSO, LISA M. (2011) *Constructing Social Research: the unity and diversity of method*, 2nd edn, Los Angeles: Sage.

RANTALA, KATI and HELLSTRÖM, EEVA (2001) "Qualitative comparative analysis and a hermeneutic approach to interview data", *International Journal of Social Research Methodology: Theory & Practice* 4(2), pp. 87–100.

RIHOUX, BENOîT and RAGIN, CHARLES C. (2009) *Configurational Comparative Methods: qualitative comparative analysis (QCA) and related techniques*, Thousand Oaks, CA: Sage.

RØDLAND, KJARTAN and ODDEKALV, KURT WILLY (2006) *Ridderne i Ludeboden: historien om Norges miljøvernforbund og Kurt Willy Oddekalv*, Bergen: Miljøforlaget.

RØKEBERG, TORIL (1996) *Medlemsfrafall i den organiserte miljøbevegelsen: en studie av Norges Naturvernforbund*, Oslo: Universitetet.

RONESS, PAUL G. (1996) *Det norske organisasjonssamfunnet: litteratur om landsomfattande frivillige organisasjonar 1945–1996*, Bergen: LOS-senteret.

SÆTEN, ANNE JORUN (1991) *Miljøstiftelsen Bellona: i spennigsfeltet mellom statlig miljøforvaltning og frivillig miljøbevegelse*, Bergen: University of Bergen.

SAMPEI, YUKI and AOYAGI-USUI, MIDORI (2009) "Mass-media Coverage, Its Influence on Public Awareness of Climate-change Issues, and Implications for Japan's National Campaign to Reduce Greenhouse Gas Emissions", *Global Environmental Change* 19(2), pp. 203–12.

SELLE, PER (1991) *Organisasjonssamfunnet i ei brytningstid: utfordringar i 90-åra*, Bergen: Alma Mater.

SØGÅRD, CHRISTINA (1995) *"Fra rebeller til konsulenter": en sosiologisk studie av miljøstiftelsen Bellona*, Bergen: University of Bergen.

STATLE, GRO (1993) *Kvinner og miljøsak: holdninger og handlinger i Norges Naturvernforbund*, Oslo: Universitetet i Oslo.

STRØMSNES, KRISTIN (2001) *Demokrati i bevegelse*, Bergen: Universitetet i Bergen.

STRØMSNES, KRISTIN (2003) "Samspillet mellom frivillig organisering og demokrati: Teoretiske argument og empirisk dokumentasjon", paper presented at the Den 11. nasjonale fagkonferanse i statsvitenskap, Trondheim.

TVEDT, TERJE (1998) *Angels of Mercy, or Development Diplomats? NGOs & foreign aid*, Trenton, NJ: Africa World Press.

TVEDT, TERJE (2002) *Verdensbilder og selvbilder. En humanitær stormakts intellektuelle historie*, Oslo: Universitetsforlaget.

TVEDT, TERJE (2003) *Utviklingshjelp, utenrikspolitikk og makt. Den norske modellen*, Oslo: Gyldendal akademisk.

WEYKER, S. (2002) "The Ironies of Information Technology", in: A. Brysk (Ed.), *Globalization and Human Rights*, Berkeley and Los Angeles: University of California Press, pp. 115–32.

WEYLER, REX (2004) *Greenpeace: an insider's account: how a group of journalists, ecologists and visionaries changed the world*, London: Rodale.

WIKER, LIV MERETE (1990) *Bellona: en studie av en pressgruppe i et offerperspektiv*, Oslo: University of Oslo.

WILLETS, PETER (1982) *Pressure Groups in the Global System: the transnational relations of issue-oriented non-governmental organizations*, London: F. Pinter.

YTTERSTAD, ANDREAS (2008) "Den balanserte, etiske aktivisten? En utforskning av Aftenposten journalistrolle i klimakampen", paper presented at the Norsk medieforskerkonferanse, Lillehammer.

TALKING POINTS AMMO
The use of neoliberal think tank fantasy themes to delegitimise scientific knowledge of climate change in Australian newspapers

Elaine McKewon

The Institute of Public Affairs (IPA) is an Australian neoliberal think tank and high-profile news source that rejects the evidence of anthropogenic climate change and opposes mitigation strategies such as an Emissions Trading Scheme (ETS). This paper uses Ernest Bormann's Symbolic Convergence Theory firstly to identify the anti-climate science fantasy themes developed by the IPA and then to trace the chaining out of these fantasy themes from the IPA into the news media. The data analysed include: (1) magazine articles published in The IPA Review during 1989–2009; (2) op-eds published by IPA senior staff in Australian newspapers during 1989–2009; and (3) editorials and opinion columns that praised IPA associate Ian Plimer and his climate "sceptic" book during April to June 2009 in the lead-up to the first Australian parliamentary debates on introducing an ETS.

Introduction

Despite the nearly universal (97 per cent) scientific consensus on the tenets of anthropogenic climate change as outlined by the Intergovernmental Panel on Climate Change (IPCC) (Anderegg et al., 2010), there is still an ongoing "scientific" debate on the reality, causes and consequences of climate change in certain sections of the Australian news media (Manne, 2011; McKewon, 2009; Taylor, 2010). This is a problem because most people get their information about science from the news media (Nelkin, 1987), especially climate science (Wilson, 1995); this includes most journalists who do not specialise in science or environment reporting (Wilson, 2000). News coverage of science shapes public understanding and action, which in turn influences the translation of science into policy (Boykoff and Rajan, 2007). Lack of public support for policies to reduce greenhouse gas emissions reduces the likelihood that governments will take effective action in time to avoid the worst-case scenarios of climate change.

As the scientific consensus on climate change strengthened over the past three decades, the news media in a number of countries created the false impression of an increasing "scientific debate"; these countries include the United States (Antilla, 2005, 2010; Boykoff and Boykoff, 2004), the United Kingdom (Boykoff and Mansfield, 2008; Carvalho, 2007) and Australia (Manne, 2011; McKewon, 2009; Taylor, 2010). Taylor's frame analysis found that the Australian news media began to diverge from the scientific consensus on climate change during the 1990s, when news framing of climate change shifted:

> from a clear and committed-to-action early understanding ... to confusion and "justified" inaction as the 1990s progressed. This occurred despite the science information about causes, effects and risks changing remarkably little and has continued beyond the study period to contemporary discourse. (2010, p. 5)

Like previous studies conducted in the United States and the United Kingdom, Taylor found that some sections of the news media give equal or greater attention and status to contrarians who are unconvinced by the evidence of anthropogenic climate change and remain opposed to mitigation strategies. Manne's (2011) analysis of news stories and opinion columns published in *The Australian* newspaper during January 2004 to April 2011 found that, of the total 880 items in the sample, only 180 accepted the scientific consensus and the need for action on climate change while 700 rejected them. Having set the objects of analysis as the news media and news media texts, most previous studies on climate change in the news media tend to provide "internalist" reasons for the divergence between the news media's reporting of climate science and the scientific consensus; these reasons include the objectivity norm in journalism (Boykoff and Boykoff, 2004; Taylor, 2010), journalists' poor understanding of the science (Antilla, 2005; Boykoff and Mansfield, 2008) and self-censorship (Antilla, 2010). However, while Boykoff and Mansfield (2008) conducted interviews with journalists to investigate their hypothesis, generally these studies have not clearly demonstrated that their conclusions are unambiguously supported by the data. Other studies have drawn more reliable conclusions regarding "internalist" reasons for the over-representation of climate contrarian views in some sections of the news media—for example, some studies have compared newspapers across the political spectrum and determined that the news organisations with conservative ideological cultures tended to favour climate "sceptics" (Carvalho, 2007; Manne, 2011; McKewon, 2009; Painter, 2011)—as the observations were drawn from the data.

Neoliberal Think Tanks as News Sources

This study expands the scope of analysis to examine "externalist" factors such as the influence exerted by news sources who seek to discredit authoritative scientific research. For the purposes of this paper, the news media are defined as the traditional publishing platforms—such as newspapers, radio and television—through which daily reports and commentary on current events are distributed to the general public. Access to the news media offers news sources a profound form of power, because the media provide the public with information that is outside their direct experience or everyday lives, thereby shaping their sense of reality about the world (Lippmann, 1922). McCombs and Shaw (1972) further propose that the news media set the agenda of topics for public thought and discussion, and play a key role in the social construction of reality (see also McCombs, 2005).

Gans describes news sources as "holders of power" who should be studied to determine which groups they represent and what "agendas [and] interests they pursue in seeking access to the news" (1979, p. 360). He further argues that "emphasising the role of sources is the best way . . . to connect the study of journalism to the larger society" (in Schlesinger, 1990, p. 61). This view is supported by Schlesinger (1990; Schlesinger and Howard, 2004), who sees news sources as "competitive definers" engaged in a contest to define the truth; this is part of the broader process whereby societal and political conflicts play out in the news media. He warns against "excessive media centrism" in journalism research (Schlesinger, 1990, p. 61) and suggests that news sources be studied as "political entrepreneurs" who aim to "influence the political agenda and shape the interpretation of current issues" (1990, p. 79). Similarly, Ericson et al. (1989) posit that the reality constructed through the news media originates with news sources, who access the media with the aim of achieving definitional advantage.

Neoliberal think tanks are among the most well-funded and influential news sources in the United States and Australia. Allen defines them as "professional social movement organisations funded by economic elites to influence public opinion and the agendas of political elites" (1992, p. 90). The primary role of neoliberal think tanks is to access the media in order to influence public opinion and advocate policies that are consistent with the interests of their organisational sponsors (Allen, 1992). Stone distinguishes the "old guard institutes" that characterised the early academic, non-political tradition of think tanks from the "new partisans" such as neoliberal think tanks which are "increasingly entrepreneurial and likely to be more . . . directly policy focused and partisan in their research and analysis" (1996, p. 18). Nevertheless, these new partisans present themselves as non-partisan policy-research institutes (Cahill, 2004; Lieberman, 2000; Smith and Marden, 2008; Soley, 1993) and operate as non-profit organisations, with all donations from sponsors tax-deductable.

In keeping with the interests of the economic elites who fund them, neoliberal think tanks promote core values of the political Right—free market capitalism, anti-socialism, privatisation, small government and deregulation (Cahill, 2004; Mirowski and Plehwe, 2009); this includes opposition to industry oversight and environmental regulations such as an ETS (Dunlap and McCright, 2010; Jacques et al., 2008; Oreskes and Conway, 2010). Their rhetorical approach to mounting opposition to environmental regulations is to oppose the science on which the proposed regulations are based.

As neoliberal think tanks are not academic or scientific organisations, their strategy for neutralising the consensus in a number of scientific fields has often involved recruiting contrarian scientists (often not experts in the relevant field) who are willing to undermine the scientific consensus in interviews with the news media; this creates the impression of a genuine "scientific debate" while legitimising attacks on authoritative scientific research (Jacques et al., 2008; Oreskes and Conway, 2010). This strategy was pioneered by the tobacco industry in the United States during the 1950s (Hertsgaard, 2006; Mirowski, 2011; Muggli et al., 2004; Union of Concerned Scientists, 2007) and has been used to undermine public confidence in scientific research that linked tobacco smoke with lung cancer, coal smoke with acid rain, and chlorofluorocarbons with stratospheric ozone depletion (Oreskes and Conway, 2010; Rampton and Stauber, 2001).

Since the 1990s, neoliberal think tanks funded by the fossil fuel, mining and energy industries in the United States have formed alliances with contrarian "experts"—many of whom have no expertise in the field of climate science—who are prepared to challenge the scientific consensus on climate change (Beder, 2002; Dunlap and McCright, 2010; Gelbspan, 2004; Jacques et al., 2008; Leggett, 2001; Oreskes and Conway, 2010); the same strategy has been used by Australian think tanks to reconstruct the scientific consensus on climate change as a "scientific debate" in the Australian news media (Hamilton, 2010, 2007; McKewon, 2009; Pearse, 2007). This is consistent with the interests of donors such as American oil billionaires Charles and David Koch (Mayer, 2010), Australian mining magnate Hugh Morgan (Smith and Marden, 2008) and the world's biggest oil company, ExxonMobil. During the years 1998 and 2004, it is estimated ExxonMobil alone donated $16 million to think tanks that ran media campaigns to "manufacture doubt" on the science of climate change (Adam, 2009; Mooney, 2005; Union of Concerned Scientists, 2007).

Neoliberal think tanks also often construct media events by publishing books that promote "environmental scepticism" in the lead-up to major environmental policy debates (Jacques et al., 2008). These constructed media events maximise news coverage at the most politically strategic time to influence public opinion and the policy agenda. Of 141 English-language "environmentally sceptical" books published between 1972 and 2005, Jacques and colleagues found that 92 per cent were published by neoliberal think tanks or their associates; meanwhile, they found that 90 per cent of neoliberal think tanks that engage in environmental issues espouse "scepticism". The majority of these books were published in the months before two major environmental conferences: the Earth Summit in Rio de Janeiro in 1992 and the third session of the Conference of Parties to the UNFCCC which adopted the Kyoto Protocol in 1997 (Jacques et al., 2008).

The Institute of Public Affairs

The IPA is an Australian neoliberal think tank and high-profile news source that rejects the evidence of anthropogenic climate change and opposes mitigation strategies such as an ETS. It has long-standing connections to the right-wing Liberal Party in Australia, having been founded by senior members of the Liberal Party in 1943. Yet the IPA describes itself as an "independent . . . public policy think tank" that promotes "the free market of ideas, the free flow of capital [and] a limited and efficient government" (IPA, 2010a).

Like its counterparts in the United States, the IPA also opposes environmental regulations. In 2005, the IPA established the Australian Environment Foundation, which is marketed as a grassroots environmental movement (*The Canberra Times*, 2005), but actually fights to prevent or reverse environmental regulations (Fyfe, 2005). In 2008, following the federal election of the (Labor) Rudd government, which promised to introduce an ETS, the Australian Environment Foundation spawned a second-tier IPA front group, the Australian Climate Science Coalition (ACSC), which consists mainly of an advisory board of contrarian scientists. Through media interviews, newspaper op-eds, magazine articles and books, the ACSC rejects the evidence of anthropogenic climate change and opposes mitigation policies (ACSC, 2009).

Along with these two front groups, the IPA is the most active and vocal Australian neoliberal think tank that uses its access to the news media to dispute anthropogenic global warming and oppose policies to reduce greenhouse gas emissions (Hamilton, 2007; Pearse, 2007, 2009). Its major corporate donors include companies in the fossil fuel, mining and energy industries (Fist, 2000; Hannan and Carney, 2005) including coal-fired electricity generators in Victoria and New South Wales (Pearse, 2009). The IPA also receives substantial donations from oil companies (Frew, 2007) and the tobacco and forestry industries (Vincent, 2004).

The IPA's source–media strategy includes publishing op-eds in newspapers, being quoted as a news source and providing "expert" commentary on radio and television (IPA, 2010b). The IPA frequently invites newspaper editors, journalists and broadcasters to its lectures, policy seminars (Cahill, 2004) and the launches of books written by IPA staff and associates (McKewon, 2009). Its quarterly magazine, *The IPA Review*, is distributed to members and sold through local newsagents (Cahill, 2004).

Right-wing newspaper columnists in Australia often look to the IPA and its front groups to articulate ideologically consistent rhetoric and policy positions on a range of environmental issues, including climate change (Pearse, 2007). Cahill (2004) also found that think tanks like the IPA form the "backbone of the neoliberal movement" and function as authors of policy ideas and rhetoric that chain out through (what Hajer, 1993 called) "discourse coalitions", which include right-wing editors and journalists: "The [neoliberal] movement has furnished journalists—particularly fellow travelling sympathisers—with the movement's rhetorical arsenal: an arsenal which has been vigorously used to demonise [its opponents]" (Cahill, 2004, p. 270).

Background—Plimer Book

In the lead-up to the first Australian parliamentary debates on introducing an ETS in 2009, Ian Plimer—mining geologist, mining company director (Ivanhoe Australia, 2009), IPA associate scholar (IPA, 2009a), ACSC scientific advisor (ACSC, 2009), and climate change contrarian—published a book that argues there is no link between human activities and climate change. Plimer boasted that his book, *Heaven and Earth: global warming—the missing science* (2009a), would "knock out every single argument we hear about climate change" (Jenkin, 2009). The book was not peer-reviewed before publication and was widely discredited by scientists on the basis that it reflects a poor understanding of the climate system, includes unattributed material (including graphs), presents misleading or incorrect data, fails to present evidence to support conclusions, misrepresents scientific studies cited as support for his claims, and relies on historical arguments that have long been refuted and abandoned in peer-reviewed scientific journals (Ashley, 2009; Brook, 2009; Enting, 2009; Glikson, 2009; Karoly, 2009; Lambeck, 2009; Morton, 2009; Sandiford, 2009; Veron, 2009; Walter, 2009; Woodroffe et al., 2009). The only scientists who gave the book positive reviews were those who, like Plimer, were associated with the IPA and its front group, the ACSC (Carter et al., 2009; Kinninmonth, 2009).

Plimer's book and the coverage it generated did not just challenge the scientific consensus on climate change; it presented an elaborate alternative reality composed of conspiracy-fuelled fantasy themes developed by the IPA; for example, "Global warming has become the secular religion of today" (2009a, pp. 462–3) and "The environmental religion embraces anti-human totalitarianism" (2009a, p. 466). Although neoliberal think tanks and right-wing politicians and pundits in the United States have been using the same fantasy themes since at least the 1990s (see for example Coffman, 1994; Nelson, 1996), the earliest references I have found of the above fantasy themes are in *The IPA Review* in 1989. The publication of these and several other IPA fantasy themes in *Heaven and Earth* enabled them to chain out to a wider audience during three months of high-profile media coverage in Australia (April to June 2009). During this time, the IPA was never mentioned, despite the book's strong links to the organisation. Apart from Plimer's direct link as an associate scholar, the IPA's Executive Director, John Roskam, sits on the editorial board of Connor Court Publishing (Connor Court Publishing, 2009a; IPA, 2009b) which published *Heaven and Earth* (Connor Court Publishing, 2009b). The IPA also launched the book and promoted it on the IPA's website (IPA, 2009c), while the publisher sent advance copies to right-wing newspaper journalists and editors:

> [T]here was no PR company involved and no sophisticated marketing strategy. Instead, Anthony Cappello, the publisher . . . hit the phone and rang his mates. The fact that he is well connected with the conservative media was his greatest asset, along with the fact that his contacts are desperate for a counter view on global warming . . . Cappello says the success [of the book] is due to all the pre-publicity he's been doing since September, although again this seems to have been targeted mostly at the right. (Dodd, 2009)

Heaven and Earth generated 219 articles in Australia's national newspapers, metropolitan dailies and regional newspapers. This coverage included news stories, feature articles, science and environmental reports, editorials, op-eds, opinion columns and letters to the editor. Most of the coverage was favourable and presented Plimer as a non-partisan, independent "expert" on climate science (McKewon, 2009).

Methodology

This paper uses a triangulation of three datasets for the purpose of analysis. The initial stage of this analysis involves the identification of the anti-climate science fantasy themes developed by the IPA and published in its quarterly magazine, *The IPA Review*. The analysis then traces the chaining out of these fantasy themes from the IPA into Australian newspapers via intertextual chains routed through two channels.

In the first (direct) channel, IPA senior staff publish op-eds that contain the fantasy themes in Australian newspapers. The op-ed represents a prestigious form of newspaper coverage for a news source: it is authored by the news source and so the source's version of reality is presented undiluted and uncrowded by the reporting of news events, editorial framing or competing news sources quoted in the same news story. At the same time, the newspaper's publication of the op-ed bestows a special status on the news source—what Fairclough (1995) might call "authorised knower".

In the second (indirect) channel, right-wing newspaper editors and columnists use the IPA's fantasy themes as the rhetorical basis for editorial leaders and opinion pieces praising Plimer and his book. The writers in this category, often senior-ranking journalists, are also presented as "authorised knowers" whose opinion-based commentary about reported news and current events is valued by the newspaper and its readers. The communicative tone employed by editors and opinion columnists is characterised by authoritative language, declarative sentences and categorical statements. This authoritative communication style also serves to project the audience as "receptive, waiting to be told, wanting to know" (Fairclough, 1995, p. 4).

This analysis uses a hybrid of Discourse Analysis (DA) and Fantasy Theme Analysis (FTA) to examine samples from the three datasets described above. Drawing on the theory of Fairclough (1995) on media discourse, the DA aims to answer three sets of questions: How is the world (events, relationships) represented? What identities are set up for those involved in the story (journalists, audiences, "third parties" referred to or interviewed)? What relationships are set up between those involved (e.g. journalist–audience, expert–audience relationships)? (Fairclough, 1995, p. 5). In an effort to answer these questions, the more detailed framework of the analysis is provided by FTA, taken from Symbolic Convergence Theory. FTA is a valuable tool for identifying, categorising and evaluating the components of the rhetorical vision revealed in the texts; however, the broader societal implications of the IPA's rhetorical vision and its chaining out into the news media are better understood when the fantasy themes are further analysed within the context of Fairclough's questions as outlined above.

Bormann's Symbolic Convergence Theory (SCT) is a general communication theory that offers an explanation for the process by which groups develop and share a symbolic reality. The process firstly involves the creation and sharing of stories, or *fantasy themes*, within small groups; the fantasy themes then chain out from person to person, and from group to group—often through the mass media—to create a shared social reality or *rhetorical vision*. The basic concept of SCT is the *fantasy theme*, which is a structured dramatic account of a real-world experience (Bormann, 1972). *Fantasy theme* is further defined as "the creative and imaginative shared interpretation of events that fulfills a group psychological or rhetorical need" (Bormann, 1985, p. 130). Fantasy themes are stories driven by *dramatis personae*—stock characters such as protagonists (i.e. heroes, victims) and antagonists (i.e. villains)—and are often grounded in the personality traits, moral codes and motives of the characters:

> When someone dramatizes an event he or she must select certain people to be the focus of the story and present them in a favorable light while selecting others to be portrayed in a more negative fashion ... Interpreting events in terms of human action allows us to assign responsibility, to praise or blame, to arouse and propitiate guilt, to hate, and to love. (Bormann, 1992, pp. 368–9)

FTA involves the identification of the fantasy themes that construct a group's rhetorical vision. It also involves the identification of associated basic concepts such as characters, plot lines, settings, sanctioning agents and symbolic cues. Sanctioning agents, which justify the drama, might be a higher power or a sacred principle such as Truth or "even a concept such as 'weapons of mass destruction'" (Duffy, 2003, p. 294). A symbolic cue might be an inside joke, code word, phrase, slogan or the name of a persona or organisation that drives a fantasy theme; the function of the cue is to trigger an emotional response such as anger, hatred, love, affection and even laughter or humour from other group members. Symbolic cues provide one indication that symbolic convergence has taken place (Bormann, 1985).

Methods

Three datasets were used in the qualitative analysis: (1) magazine articles published in *The IPA Review* during 1989–2009; (2) op-eds published by IPA senior staff in Australian newspapers during 1989–2009; and (3) editorials and opinion columns that gave favourable coverage to Ian Plimer and his book during April to June 2009 in the lead-up to the first Australian parliamentary debates on introducing an ETS.

The first sample of magazine articles was generated by searching the IPA website's archives of *The IPA Review* including the years 1989–2009, as 1989 is the year in which *The IPA Review* published its first article on the enhanced Greenhouse Effect (Roskam, 2010). It is also the year before the IPCC released its First Assessment Report on the science of climate change. A total of 67 *IPA Review* editorials and articles were found that relate to climate science. The items selected for analysis were those written by IPA senior staff and associates who explicitly articulated the recurring fantasy themes relating to climate science.

The second sample of newspaper op-eds was generated through the Dow Jones Factiva database using the date range 1 January 1989 to 30 June 2009; as 1989 is the year in which the IPA took up the issue of global warming and June 2009 marks the end of peak Australian newspaper coverage of Ian Plimer's book. The newspapers searched include Australia's national and metropolitan daily newspapers including weekend editions. The first tier of search terms included "greenhouse effect" or "global warming" or "climate change". This first tier was paired with a second tier of search terms that included "Institute of Public Affairs" or "Australian Environment Foundation" or "Australian Climate Science Coalition". There being no filter for targeting only op-eds, the sample of 229 items included news stories, feature articles, science and environmental reports, editorials, op-eds, opinion columns and letters to the editor. The items selected for analysis were climate science-related op-eds in which the IPA's fantasy themes identified in the first dataset were most clearly re-animated.

The third sample of editorials and opinion columns was generated through the Dow Jones Factiva database using the date range 1 April 2009 to 30 June 2009 and the search term "Ian Plimer". The newspapers searched include Australia's national, metropolitan daily and regional newspapers, including weekend editions. There being no filter for targeting only editorials and opinion columns, this returned a total of 219 items including news stories, feature articles, science and environmental reports, editorials, op-eds, opinion columns and letters to the editor. The items selected for analysis were editorials and opinion columns published during the two weeks immediately surrounding the launch of *Heaven and Earth* as well as two newspaper op-eds published by Plimer later in the study period.

Results

Nine discrete fantasy themes were identified across all three datasets. These fantasy themes were grouped into two classifications: "The Plea for Scientific Truth" and "Religious, Political and Economic Conspiracies". The underlying claim for the first group of fantasy themes is that the climate "sceptics" are right—that the Earth is not warming and that, even if it is, it is not attributable to human activities and not a serious threat. The sanctioning agent here is Truth. For the second group of fantasy themes, the uniting claim is that the threat of climate change has been deliberately fabricated or exaggerated in order to enable various groups to advance hidden agendas—thus, the sanctioning agent for these fantasy themes is Integrity. These two categories will be discussed first in the sections below, followed by a more detailed analysis of three of the more prominent fantasy themes found in the data.

The Plea for Scientific Truth

There are four fantasy themes in this category: "Climate Scientists as Rent-seeking Frauds"; "Climate Scientists as Dissent-stifling Elite"; "Plimer as Galileo"[1] and "Plimer as the People's Scientist". In terms of *dramatis personae*, the villains in these fantasy themes are "elitist" climate scientists, especially those associated with the IPCC. The supporting cast of villains includes government agencies that fund the "propaganda" produced by the scientific elite. The heroes are contrarian or "sceptic" scientists who reject the scientific consensus and Speak Truth to Power at the risk of incurring the wrath of the iron-fisted establishment.

These fantasy themes tell the story of a global cabal of climate scientists who are consumed with protecting their privileged status and blind to the reality that the scientific consensus on climate science has no evidentiary basis. The primary plot line sees a powerful scientific elite dominating and controlling the field of climate science and suppressing the Scientific Truth by persecuting the scientific voices of dissent.

Religious, Political and Economic Conspiracies

There are six fantasy themes in this category: "Climate Scientists as Rent-seeking Frauds"; "Climate Science as Religion"; "Environmentalism as Religion"; "Climate Science as Left-wing Political Conspiracy"; Green as the New Red" and "Climate Change Mitigation as Money-spinning Scam". The fantasy theme "Climate Scientists as Rent-seeking Frauds" appears in both groupings because it has separate components that speak to both categories. The *dramatis personae* in these fantasy themes are all cast as villains—climate scientists, the United Nations, Al Gore, the political Left, environmental activists, investors in the renewable energy industry and carbon traders.

The three fantasy themes that relate directly to climate science are discussed here. According to the plot line of the fantasy theme "Climate Scientists as Rent-seeking Frauds", climate scientists are not engaged in a pursuit of the truth through science—on the contrary, climate scientists routinely fabricate data, misrepresent the results of their studies and exaggerate the threat of climate change to extract lucrative research grants from the government.

The fantasy theme "Climate Science as Religion" posits that climate science is *not science*—it is a manifestation of the new "environmental religion". This fantasy theme enables evidence-based scientific conclusions to be dismissed as an arbitrary set of beliefs or dogma. The plot line of the fantasy theme "Climate Science as Left-wing Political Conspiracy" sees the environmental religion's leftist allies (Labor and Green political parties, and even the United Nations) using climate change as a "scare tactic" to consolidate their political power, increase taxes to redistribute wealth, and impose a New World Order that will compromise national sovereignty and restrict personal freedoms. These two fantasy themes serve to delegitimate the most vocal social groups who accept the scientific consensus and support action on climate change: the environmental movement and the political Left. They are portrayed not as people rationally responding to a real environmental threat identified by the science: they are variously cast as irrational religious fundamentalists following a doomsday cult or as left-wing conspirators cynically using a fabricated or exaggerated threat to pursue political goals.

The following two sections offer examples from each dataset to illustrate three of the most prominent fantasy themes in greater detail: "Climate Science as Religion"; "Environmentalism as Religion" and "Climate Scientists as Rent-seeking Frauds".

"Climate Science as Religion"/"Environmentalism as Religion"

The excerpts below, taken from all three datasets, are examples of how the "Climate Science as Religion" and "Environmentalism as Religion" fantasy themes are articulated.

Although these are two discrete fantasy themes, it was often not possible to separate them because time and again the rhetoric conflates climate scientists with the environmental movement and climate science with religion. It is also not unusual for other fantasy themes to be supplemented within the same excerpt, since all the fantasy themes combined serve to construct the overarching rhetorical vision. Symbolic cues for these two religious fantasy themes include words such as "religion", "faith", "hymn", "doom" and "fundamentalism".

The IPA's hostility towards the increasing scientific consensus on anthropogenic climate change is documented in *The IPA Review* from the year 1989, in the year leading up to the IPCC's First Assessment Report which reviewed the most recent climate science research. In October 1989, *The IPA Review* published an unattributed editorial entitled "The Green Messiah" which suggests that concerns about the enhanced Greenhouse Effect are overstated and based not on a scientifically identified problem, but on religious motivations. The argument is made that the "Green Messiah" (environmentalism) and its by-products such as climate science are filling the voids left by the decline of traditional religion:

> [W]ith the declining capacity of the traditional churches to satisfy the religious yearnings of people ... such yearnings are seeking other outlets. The birth of a modern form of pantheism (nature worship) in the rise of the environmental movement may be one such outlet ... The problem is to decide where science ends and faith begins: how seriously should we take the prophets of environmental doom? (IPA, 1989, p. 4)

The editorial is also the earliest example of the IPA's conflation of climate scientists with environmental activists. One clear example of this combined fantasy theme chaining out is an op-ed entitled "Sceptics, the Environment Needs You", published by Jennifer Marohasy (a biologist and former director of the IPA front group, the Australian Environment Foundation) in *The Age* newspaper in 2004. Marohasy labels the environmental movement a "new religion" of "environmental doomsayers" whose concerns are not based on scientific evidence:

> [E]nvironmentalism is emerging as a new religion, with Greenpeace, the World Wide Fund for Nature and the World Wildlife Fund representing the new church ... let us move forward on the basis of the evidence, rather than the failed predictions of our environmental doomsayers. (Marohasy, 2004)

The IPA's religious fantasy themes continued to chain out via editorials and opinion columns when IPA associate Plimer's book was launched in April 2009. *The Weekend Australian* conservative newspaper ran an editorial under the headline "More Heat Than Light" which argues that concerns about climate change are not based on science but the zealotry of environmental activists. With its clear religious and persecutory overtones—for example, its use of the word "heresy'—the editorial hints at the "Plimer as Galileo" fantasy theme as well. It also explicitly reconstructs climate science as a "faith-based" rather than an evidence-based pursuit:

> For environmental activists, any suggestion prophecies of planetary peril should be considered carefully is heresy ... to assume we know how much the planet will warm this century and what effect this will have is a matter of faith, not reason. And faith-based research is less science than secular religion. (*The Australian*, 2009b)

This editorial also directs readers to an opinion piece written by the paper's Adelaide bureau chief, Jamie Walker, who favourably reviewed Plimer's book.

In this 1600-word opinion piece entitled "The Climate of Consensus is Precipitating a Disaster for Science", Walker (2009) reduces the scientific consensus on anthropogenic climate change to a mere "assumption" that forms the basis of "catastrophic consequences talked up by the doom-and-gloom merchants". He also conjures a religious image of "[then Australian Prime Minister] Kevin Rudd, [paleontologist] Tim Flannery and former US vice-president Al Gore singing from the same hymn sheet about the hydra-headed menace of global warming". Walker then clearly invokes the "Climate Science as Religion" fantasy theme by quoting Plimer's book—"Global warming has become the secular religion of today"—and including excerpts from his interview with Plimer, who conflates climate scientists and environmental activists into a single category of "religious fundamentalists" akin to creationists:

> "The creationists were trying to teach a religious fundamentalism dressed up as science... and they totally changed the nature science," [Plimer] says ... "So I make great comparison between the way creationists operate and the way some of the rabid environmentalists and global warmers operate." (Plimer, cited in Walker, 2009)

A few pages away in the same edition of *The Weekend Australian*, opinion columnist Christopher Pearson wrote that he would be "honoured to serve as the master of ceremonies" at the Adelaide launch of Plimer's book the following week. In his 1300-word review of Plimer's book, "Sceptic Spells Doom for Alarmist Religion", Pearson (2009) said that accepting the evidence of anthropogenic climate change "requires many non-scientific leaps of faith" and that the IPCC's stated 90 per cent ("very likely") level of certainty regarding the main driver of climate change is "comparable to 100 per cent certainty professed by religious devotees that theirs is the one and only true faith". Pearson also quotes Plimer's book, which references both religious fantasy themes and again conflates climate science with religion and scientists with activists:

> With some rabid environmentalists, human-induced global warming has evolved into a religious belief system ... The new environmental religion embraces anti-human totalitarianism. (Plimer, in Pearson, 2009)

On the same day, in the more politically progressive *Sydney Morning Herald* newspaper, conservative opinion columnist Miranda Devine's favourable review of Plimer's book ran under the headline "Planet Doomsayers Need a Cold Shower" (Devine, 2009). She says that "climate change has become a quasi-religious belief" and praises Plimer's book as a "comprehensive scientific refutation of the *beliefs* underpinning the idea of human-caused climate change" (emphasis added).

On 29 May 2009, at the peak of news coverage of his book in Australia, Plimer himself published an op-ed in *The Australian* newspaper. Drawing on the strong opposition to the Rudd government's proposed ETS in Australia's regional areas, Plimer argues that climate science and mitigation strategies are being promoted by city-based "atheist" religious "fundamentalists':

> [T]here are a large number of punters who object to being treated dismissively as stupid, who do not like being told what to think, who value independence, who resile from

> personal attacks and have life experiences very different from the urban environmental atheists attempting to impose a new fundamentalist religion. (Plimer, 2009b)

"Climate Scientists as Rent-seeking Frauds"

The following excerpts give examples of how the "Climate Scientists as Rent-seeking Frauds" fantasy theme is articulated in all three datasets; this includes manifestations of this central fantasy theme in both categories. Like most fantasy themes, it often appears in conjunction with others, especially the other three fantasy themes in the "Plea for Scientific Truth" category: "Climate Scientists as Dissent-stifling Elite", "Plimer as Galileo" and "Plimer as the People's Scientist". The symbolic cues for this category of fantasy themes include references to dissident scientists "beginning to speak out" on the "facts of climate change" to expose the "scam" or the "racket" of climate science and the "alarmism", "hype" and "hysteria" it has spawned.

The IPA began to directly attack the integrity of climate scientists following the IPCC's Second Assessment Report which states, "The balance of evidence suggests a discernible human influence on global climate" (IPCC, 1995, p. 22). In one op-ed published in *The Courier-Mail* in 2000, Ron Brunton (an anthropologist and former head of the IPA's Environmental Policy Unit) combines the tactic of conflating the environmental movement and climate scientists with allegations of rent-seeking behaviour. In this op-ed, Brunton accuses climate scientists (among others) of "cashing in" on the "new industry" that had been created by stoking the public's fears about the potential consequences of global warming:

> A new global industry has been built around the greenhouse effect, as scientists, environmentalists, bureaucrats and others seek to cash in on public fears about global warming. (Brunton, 2000)

The IPA's attacks on climate scientists and the scientific consensus intensified following the IPCC's Third Assessment Report which states, "There is new and stronger evidence that most of the warming observed over the last 50 years is attributable to human activities" (IPCC, 2001, p. 10). In 2003, *The IPA Review* published an article entitled "Science is Not Consensus" by Bob Carter, a marine geologist and founding member of the IPA front groups, the Australian Environment Foundation (AEF) and the Australian Climate Science Coalition (ACSC). In this article, Carter joins forces with American sceptics who were escalating their attacks on climate scientists and the scientific consensus; here, he quotes a former President of the Mineralogical Society of America: "The idea that humans have significantly enhanced global warming is by far the most massive abuse of science that I have ever seen" (Ross, in Carter, 2003, p. 13). Carter's own inflammatory rhetoric compares scientists to "sex workers" and accuses the entire field of climate scientists of fabricating environmental threats to procure government funding:

> To capture government's attention, and funding, requires the generation of a crisis in one of these politically sensitive areas . . . Thus did our venerable handmaiden, science, become a sex-worker. (Carter, 2003, p. 12)

Carter also returns to the IPA rhetorical tactic of conflating scientific organisations and environmental groups into a single irrational, quasi-religious or quasi-political entity: "The alternative to a scientific approach is one based on superstition, phobia, religion or politics". His claims are supported by (former) IPA Executive Director Mike Nahan. In his editorial, "The Demise of Science", published in the same issue of *The IPA Review*, Nahan agrees with Carter that climate scientists have become "myth-makers" who have abandoned the pursuit of the truth for the pursuit of funding:

> The question is: why have so many scientists also succumbed to being myth-makers? One answer is money. Shock and horror not only sells papers and generates donations for NGOs, it also generates funding for research. There has been a discernible decline in the willingness of scientists to stand up for truth and against the populist misuse of science. (2003, p. 2)

Once the IPA had established the ACSC in 2008 (in response to the Rudd government's ETS proposal), ACSC adviser Carter published an op-ed in *The Age* newspaper warning that the federal government's advisory channels had become "clogged with rent seekers, special pleaders and green activists" (Carter, 2008). The following year, another ACSC science adviser, John McLean (an Information Technology specialist), further advanced the "Climate Scientists as Rent-seeking Frauds" fantasy theme in *The Australian* newspaper during peak coverage of Plimer's book. In his op-ed entitled "Science a Slave to Expediency", McLean insisted that mainstream climate science could no longer be trusted because it was dominated by self-serving, opportunistic rent-seekers:

> Vested interests now dominate climate science ... they all wish to retain an income stream and whatever reputations they've established ... Climate science is no longer an impartial truth but a slave to the yoke of politics and opportunism. (McLean, 2009)

In the days leading up to the launch of Plimer's book, *The Australian* used very similar language in an editorial published under the headline "Global Warming Beat-ups Threaten Scientific Credibility" (2009a). This editorial said that climate science had become a "racket" dominated by a powerful, self-serving elite that silences "serious scientists", alluding to both the "Climate Science as Rent-seeking Frauds" and the "Plimer as Galileo" fantasy themes:

> Researchers who have staked their reputations and funding on human-induced global warming say the science is settled, presumably because they have the numbers to shout down naysayers ... While serious scientists are obliged to report what their research shows, whoever it offends, claims of climate change have become a racket. (*The Australian*, 2009a)

The following week, opinion columnist Christopher Pearson invoked the "Climate Scientists as Rent-seeking Frauds" fantasy theme in his favourable review of Plimer's book. Pearson even suggests that, not only is there no global warming and thus no related environmental threat, but that there is no actual academic field of climate science, only a constellation of "vested interests" and "rent seekers" devoted to generating long-term income:

> Then there are the burgeoning numbers of so-called climate scientists, all with research grants and vested interests in defending what they've come to think of as mainstream climatology. Finally there are all the rent-seekers in the academy, in government departments and in commerce who expect to make a living out of regulating carbon emissions. (Pearson, 2009)

On 5 May 2009, Plimer published an op-ed in *The Australian* newspaper entitled "Hot-air doomsayers" (2009b) which invoked the "Climate Scientists as Rent-seeking

Frauds" and the "Plimer as the People's Scientist" fantasy themes. Here he attempts to drive a wedge between climate scientists and the rest of society by projecting a sense of "helplessness" and "disenfranchisement" on the general public, whom he says knows they are being "conned" and "frightened witless" by elitist scientists and moralistic activists and politicians:

> My correspondents feel helpless and disenfranchised with the unending negative moralistic cacophony about climate change. They know it smells but they cannot find where the smell comes from . . . the average person knows that they are being conned and finally they have a source reference. (Plimer, 2009b)

"Plimer as the People's Scientist" also alludes to the "Plimer as Galileo" fantasy theme, only this time the hero is persecuted by a communist regime, as Plimer sensationally suggests that he could face trial for his scientific "crimes":

> I await the establishment of a Stalinist-type Truth and Retribution Commission to try me for my crimes against the established order and politicised science. (Plimer, 2009b)

Discussion and Conclusions

This analysis shows that the Institute of Public Affairs, an Australian neoliberal think tank, has used its access to the media as a news source to influence the public discussion on climate change in Australia. The rhetorical vision expressed by the IPA is one of hostility towards climate scientists and the scientific consensus on climate change; the IPA's hostility is based on its opposition to proposed government regulations aimed at reducing greenhouse gas emissions.

The distinct fantasy themes identified in this analysis were traced as they chained out from *The IPA Review* to op-eds published by IPA senior staff in major newspapers and through opinion columns written by conservative members of the press. Australian conservative columnists continue to take their lead on the issue of climate change from the IPA (Pearse, 2007), readily repeating the IPA's fantasy themes. This study also supports the conclusion of Carvalho (2007) that ideology is a primary factor that determines the news media's representation of scientific knowledge about climate change and the selection of "experts" who are given a voice. Plimer received by far the most favourable press from the editors of conservative newspapers and conservative opinion columnists, while the most unfavourable coverage originated with scientists and science and environment reporters.

The IPA's fantasy themes contained in a book written by an IPA associate scholar were given high-profile media coverage in major Australian newspapers in the lead-up to the historic debate in the Australian parliament about introducing an ETS; this is consistent with the conclusion of Jacques et al. (2008) that neoliberal think tanks tend to launch environmentally "sceptical" books in the lead-up to major debates on environmental policy, in an effort to influence the policy process.

Even the IPA's associate scholars such as Plimer and Carter devote a considerable proportion of their *IPA Review* articles, op-eds, media interviews and books to deploying the IPA's fantasy themes, or "rhetorical arsenal" (Cahill, 2004, p. 270), which they use to vilify, demonise and delegitimate climate scientists and other citizens who accept the scientific consensus and the need for action on climate change. Fantasy Theme Analysis provides an especially useful structure for identifying and examining the distinctive "talking points" used by neoliberal think tanks and their associates in their efforts to delegitimate climate scientists. In particular, the fantasy themes are key to answering the three main questions posed by Fairclough (1995) to discourse analysts.

Firstly, how is *reality* represented in the text? The fantasy themes dramatise the IPA's interpretation of reality, in which communism and the Judeo-Christian tradition have been replaced in the West with the irrational, quasi-religious environmental movement that advances a leftist political agenda of government intervention. In this reality, climate scientists have been co-opted into the "religious" order and now form a powerful, self-serving elite that does not tolerate dissenting views. This interpretation of reality enables the IPA to argue that its contrarian views on climate change are correct yet silenced by a corrupt establishment—their associate scholars are not scientifically irrelevant, they are modern-day Galileos exposing the truth about the fabrication of anthropogenic climate change, which is fuelling a money-spinning scam in the form of higher taxes and the trading of carbon credits.

Secondly, what *identities* are set up for those involved in the story? Fantasy themes use stock characters such as villains, heroes and victims to construct dramatic narratives. Climate scientists are clearly the villains: rent-seeking, dissent-stifling, Ivory Tower elitists who readily commit scientific fraud for financial gain. The fantasy themes also repeatedly conflate climate scientists and the environmental lobby—thus reconstructing the identity of climate scientists as environmental activists. Boykoff and Mansfield (2008) found that conservative tabloid opinion columnists in the United Kingdom use similar rhetorical tactics. Meanwhile, contrarian scientists are cast as the heroes who risk persecution, like Galileo, to expose the flaws in the prevailing orthodoxy. Finally, the general public are cast as "helpless and disenfranchised" victims blinded by the science and forced to bear the burden of government policies to reduce greenhouse gas emissions through lost jobs and financial hardship.

Thirdly, what *relationships* are set up between those involved? The fantasy themes construct two distinct relationships of unequal power, and a third relationship of empowerment. In the first instance, there is the adversarial relationship set up between the scientific "elite" and the "dissidents" they "persecute", with climate scientists commanding the field of climate science by monopolising government research funding and colluding with their peers to keep the work of dissident scientists out of prestigious journals. In the second instance, there is the exploitative relationship set up between the "rent-seeking Ivory Tower elites" and ordinary citizens who fund research that is based on "scientific fraud". In this relationship, the general public are "conned" and "frightened witless" by fabricated threats and forced to endure financial hardship resulting from policy based on religious and ideological propaganda. This contrasts sharply with the third relationship of empowerment set up by the fantasy themes between ordinary citizens and contrarian scientists such as Plimer, the People's Scientist, who stands up to the oppressive establishment and offers a truth-telling accessible book to the "punters" which contains "all the scientific ammunition they could want" (Pearson, 2009).

Finally, there are important social identities and relationships inherent in the production and consumption of media texts. The IPA functions as the font of a coherent rhetorical vision for ideologically conservative editors and columnists, who then present themselves as figures of authority to their readers—opinion columnists are people who have "the facts" and the "right to tell". Their authoritative, lecturing style of communication enhances their authority and establishes a relationship of unequal power—nevertheless one of trust—with their readers, who are assumed to be "receptive, waiting to be told, wanting to know" (Fairclough, 1995, p. 4). Thus, newspaper editorials and opinion columns which use the IPA's fantasy themes increase the likelihood that readers will form a distrust of climate scientists, climate science, policies to reduce greenhouse gas emissions and citizens who support those policies. It is a process that is likely to deepen the existing partisan divide over climate policy and climate science.

This study demonstrates the value of studying news sources as "holders of power" (Gans, 1979) and as "political entrepreneurs" (Schlesinger, 1990). This study also shows that analysing news sources illuminates how wider societal and political conflicts are played out in the news media.

ACKNOWLEDGEMENTS

I would like to thank my PhD supervisors, Prof. Wendy Bacon and Dr Catriona Bonfiglioli, for their constant support and guidance, and Mr Ian Grant, for his thoughtful comments on various drafts of this manuscript. An early version of this paper was presented at the ECREA pre-conference "Communicating Climate Change" at the University of Hamburg on 12 October 2010. This research was made possible by an Australian Postgraduate Award scholarship.

NOTE

1. This fantasy theme has chained out in Australia to the extent that all contrarian scientists are being portrayed as modern-day Galileos by specialist think tanks. See, for example, http://www.galileomovement.com.au/.

REFERENCES

ADAM, DAVID (2009) "ExxonMobil Continuing to Fund Climate Sceptic Groups, Records Show", *The Guardian*, 1 July, http://www.guardian.co.uk/environment/2009/jul/01/exxon-mobil-climate-change-sceptics-funding, accessed 29 January 2011.

ALLEN, MICHAEL P. (1992) "Elite Social Movement Organisations and the State: the rise of the conservative policy-planning network", *Research in Politics and Society* 4, pp. 87–109.

ANDEREGG, WILLIAM R. L., PRALL, JAMES W., HAROLD, JACOB and SCHNEIDER, STEPHEN H. (2010) "Expert Credibility in Climate Change", *Proceedings of the National Academy of Sciences of the United States of America* 107(27), pp. 12107–9.

ANTILLA, LIISA (2005) "Climate of Scepticism: US newspaper coverage of the science of climate change", *Global Environmental Change* 15, pp. 338–52.

ANTILLA, LIISA (2010) "Self-censorship and Science: a geographical review of media coverage of climate tipping points", *Public Understanding of Science* 19(2), pp. 240–56.

ASHLEY, MICHAEL (2009) "No Science in Plimer's Primer", *The Weekend Australian*, 9 May, http://www.theaustralian.com.au/news/ian-plimer-heaven-and-earth/story-e6frg8no-1225710387147, accessed 23 June 2009.

AUSTRALIAN CLIMATE SCIENCE COALITION (ACSC) (2009) "About Us", http://www.auscsc.org.au/about_us.html, accessed 23 August 2009.

BEDER, SHARON (2002) *Global Spin: the corporate assault on environmentalism*, Totnes: Green Books & Chelsea Green Publishing.

BORMANN, ERNEST G. (1972) "Fantasy and Rhetorical Vision: the rhetorical criticism of social reality", *Quarterly Journal of Speech* 58(4), pp. 396–407.

BORMANN, ERNEST G. (1985) "Symbolic Convergence Theory: a communication formulation", *Journal of Communication* 35(4), pp. 128–38.

BORMANN, ERNEST G. (1992) "Fantasy Theme Analysis and Rhetorical Theory", in: James Golden, Goodwin Berquist and William Coleman (Eds), *The Rhetoric of Western Thought*, 5th Edn, Dubuque, Iowa: Kendall/Hunt Publishing, pp. 365–84.

BOYKOFF, MAXWELL T. and BOYKOFF, JULES M. (2004) "Balance as Bias: global warming and the US prestige press", *Global Environmental Change* 14, pp. 125–36.

BOYKOFF, MAXWELL T. and MANSFIELD, MARIA (2008) "'Ye Olde Hot Aire': reporting on human contributions to climate change in the UK tabloid press", *Environmental Research Letters* 3, pp. 1–8.

BOYKOFF, MAXWELL T. and RAJAN, S. RAVI (2007) "Signals and Noise: mass media coverage of climate change in the UK and the USA", *EMBO Reports* 8(3), pp. 207–11.

BROOK, BARRY (2009) "Review: Ian Plimer's Heaven and Earth", *Crikey*, 24 April, http://www.crikey.com.au/2009/04/24/review-ian-plimers-heaven-and-earth/, accessed 3 July 2009.

BRUNTON, RON (2000) "Hidden Agenda in Petrol Protests", *The Courier-Mail*, 30 September.

CAHILL, DAMIEN C. (2004) "The Radical Neo-liberal Movement as a Hegemonic Force in Australia, 1976–1996", PhD thesis, University of Wollongong, NSW.

CARTER, BOB (2003) "Science Is Not Consensus", *The IPA Review* 55(4), pp. 11–3.

CARTER, BOB (2008) "Wong's Climate Paper Clouded With Mistakes", *The Age*, 29 July, p. 12.

CARTER, BOB, EVANS, DAVID, FRANKS, STEWART and KINNINMONTH, WILLIAM (2009) "Wong's Silent Treatment Clouds Emissions Credibility", *The Australian*, 19 June, p. 12.

CARVALHO, ANABELA (2007) "Ideological Cultures and Media Discourses on Scientific Knowledge: re-reading news on climate change", *Public Understanding of Science* 16, pp. 223–43.

COFFMAN, MICHAEL S. (1994) *Saviors of the Earth? The politics and religion of the environmental movement*, Chicago: Northfield Publishing/Moody Publishers.

CONNOR COURT PUBLISHING (2009a) "Who Is in [*sic*] Connor Court Editorial Board?", http://connorcourtpublishing.blogspot.com/2009/08/who-is-in-connor-court-editorial-board.html, accessed 5 November 2009.

CONNOR COURT PUBLISHING (2009b) "Catalog: climate change", http://www.connorcourt.com/catalog1/index.php?main_page=index&cPath=29, accessed 5 November 2009.

DEVINE, MIRANDA (2009) "Planet Doomsayers Need a Cold Shower", *The Sydney Morning Herald*, 18 April, p. 9.

DODD, ANDREW (2009) "Plimer's Heaven and Earth: a conservative coup?", *Crikey*, 20 April, http://www.crikey.com.au/2009/04/20/plimers-heaven-and-earth-a-conservative-coup/, accessed 20 April 2009.

DUFFY, MARGARET (2003) "Web of Hate: a Fantasy Theme Analysis of the rhetorical vision of hate groups online", *Journal of Communication Inquiry* 27(3), pp. 291–312.

DUNLAP, RILEY E. and MCCRIGHT, AARON M. (2010) "Climate Change Denial: sources, actors and strategies", in: Constance Lever-Tracy (Ed.), *Routledge Handbook of Climate Change and Society*, Abingdon: Routledge, pp. 240–59.

ENTING, IAN (2009) *Ian Plimer's Heaven + Earth—checking the claims*, "Australian Research Council's Centre of Excellence for Mathematics and Statistics of Complex Systems", http://www.complex.org.au/tiki-download_file.php?fileld=91, accessed 25 October 2009.

ERICSON, RICHARD V., BARANEK, PATRICIA M. and CHAN, JANET B.L. (1989) *Negotiating Control: a study of news sources*, Toronto: University of Toronto Press.

FAIRCLOUGH, NORMAN (1995) *Media Discourse*, London: Hodder Education Publishing.

FIST, STEWART (2000) "Science Fraud in the Spotlight", *The Australian*, 1 August, p. 60.

FREW, WENDY (2007) "Minchin Denies Climate Change Man-made", *The Sydney Morning Herald*, 15 March, p. 3.

FYFE, MELISSA (2005) "New Green Group Makes Conservationists See Red", *The Age*, 8 June, p. 3.

GANS, HERBERT J. (1979) *Deciding What's News: a study of CBS Evening News, NBC Nightly News, Newsweek and Time*, New York: Vintage Books.

GELBSPAN, ROSS (2004) *Boiling Point: how politicians, big oil and coal, journalists, and activists are fueling the climate crisis—and what we can do to avoid disaster*, New York: Basic Books.

GLIKSON, ANDREW (2009) "Plimer Wants to Talk Science? OK, here goes...", *Crikey*, 5 May, http://www.crikey.com.au/2009/05/05/plimer-wants-to-talk-science-ok-here-goes/, accessed 2 July 2009.

HAJER, MAARTEN A. (1993) "Discourse Coalitions and the Institutionalization of Practice: the case of acid rain in Great Britain", in: Frank Fischer and John Forester (Eds), *The Argumentative Turn in Policy Analysis and Planning*, London: University College London Press.

HAMILTON, CLIVE (2007) *Scorcher: the dirty politics of climate change*, Melbourne: Black Inc.

HAMILTON, CLIVE (2010) *Requiem for a Species: why we resist the truth about climate change*, London: Earthscan.

HANNAN, EWIN and CARNEY, SHAUN (2005) "Thinkers of Influence", *The Age*, 10 December, p. 6.

HERTSGAARD, MARK (2006) "While Washington Slept", http://www.vanityfair.com/politics/features/2006/05/warming200605, accessed 15 October 2009.

INSTITUTE OF PUBLIC AFFAIRS (IPA) (1989) "The Green Messiah", *The IPA Review* 43(2), p. 4.

INSTITUTE OF PUBLIC AFFAIRS (IPA) (2009a) "People and Associates", http://www.ipa.org.au/people/ian-plimer, accessed 23 August 2009.

INSTITUTE OF PUBLIC AFFAIRS (IPA) (2009b) "People and Associates", http://www.ipa.org.au/people/john-roskam, accessed 5 November 2009.

INSTITUTE OF PUBLIC AFFAIRS (IPA) (2009c) "Controversial Bestselling Book on Climate Change to Be Launched in Melbourne", http://www.ipa.org.au/publications/1679/controversial-bestselling-book-on-climate-change-to-be-launched-in-melbourne, accessed 18 October 2009.

INSTITUTE OF PUBLIC AFFAIRS (IPA) (2010a) "About the Institute of Public Affairs", http://www.ipa.org.au/about, accessed 15 September 2010.

INSTITUTE OF PUBLIC AFFAIRS (IPA) (2010b) "Financial Report for the Year Ended 30 June 2010", http://www.ipa.org.au/about, accessed 29 January 2011.

INTERGOVERNMENTAL PANEL ON CLIMATE CHANGE (1995) *IPCC Second Assessment: Climate Change 1995 – A Report of the Intergovernmental Panel on Climate Change*, Cambridge: Cambridge University Press.

INTERGOVERNMENTAL PANEL ON CLIMATE CHANGE (2001) *Climate Change 2001: The Scientific Basis. Contribution of Working Group I to the Third Assessment Report of the Intergovernmental Panel on Climate Change*, J. T. Houghton, Y. Ding, D. J. Griggs, M. Noguer, P. J. Van Der Linden, X. Dai, K. Maskell and C. A. Johnson (Eds), Cambridge and New York: Cambridge University Press.

IVANHOE AUSTRALIA (2009) "Board of Directors", http://www.ivanhoeaustralia.com/s/Board_of_Directors.asp, accessed 28 August 2009.

JACQUES, PETER J., DUNLAP, RILEY E. and FREEMAN, MARK (2008) "The Organisation of Denial: conservative think tanks and environmental scepticism", *Environmental Politics* 17(3), pp. 349–85.

JENKIN, CARLA (2009) "Humans Not Hurting the Climate", *The Advertiser*, 13 April, p. 10.

KAROLY, DAVID (2009) "Heaven and Earth—review by David Karoly", *The Science Show*, Australian Broadcasting Corporation, Radio National, 13 June, http://www.abc.net.au/rn/scienceshow/stories/2009/2593166.htm, accessed 8 September 2009.

KINNINMONTH, WILLIAM (2009) "Cold, Hard Facts Dispel Theories on Warming", *The Australian*, 29 April, p. 12.

LAMBECK, KURT (2009) "Comments on Heaven and Earth: Global Warming—The Missing Science", *Ockham's Razor*, Australian Broadcasting Corporation, Radio National, 7 June, http://www.abc.net.au/rn/ockhamsrazor/stories/2009/2589206.htm, accessed 8 September 2009.

LEGGETT, JEREMY (2001) *The Carbon War: global warming and the end of the oil era*, New York: Routledge.

LIEBERMAN, TRUDY (2000) *Slanting the Story: the forces that shape the news*, New York: New Press.

LIPPMANN, WALTER (1922) *Public Opinion*, New York: Macmillan.

MANNE, ROBERT (2011) "Bad News: Murdoch's *Australian* and the shaping of the nation", *Australian Quarterly Essay*, No. 43, Melbourne: Black Inc.

MAROHASY, JENNIFER (2004) "Sceptics, the Environment Needs You", *The Age*, 11 February, http://www.theage.com.au/articles/2004/02/10/1076388362550.html, accessed 20 January 2011.

MAYER, JANE (2010) "The Billionaire Brothers Who Are Waging a War Against Obama", *The New Yorker*, 30 August, http://www.newyorker.com/reporting/2010/08/30/100830fa_fact_mayer?printable=true, accessed 23 September 2010.

MCCOMBS, MAXWELL E. (2005) *Setting the Agenda: the mass media and public opinion*, Cambridge: Polity Press.

MCCOMBS, MAXWELL E. and SHAW, DONALD L. (1972) "The Agenda-setting Function of Mass Media", *Public Opinion Quarterly* 36(2), pp. 176–87.

MCKEWON, ELAINE (2009) "Resurrecting the War-by-media on Climate Science: Ian Plimer's Heaven + Earth", in: *Refereed Proceedings of the 2009 JEA Conference: Journalism education in the digital age*, Journalism Education Association of Australia, http://www.2009jeaconference.au.com/papers.html.

MCLEAN, JOHN (2009) "Science a Slave to Expediency", *The Australian*, 6 May, p. 16.

MIROWSKI, PHILIP (2011) *Science Mart: privatizing American science*, Cambridge, MA: Harvard University Press.

MIROWSKI, PHILIP and PLEHWE, DIETER (2009) *The Road from Mont Pelerin: the making of the neoliberal thought collective*, Cambridge, MA: Harvard University Press.

MOONEY, CHRIS (2005) "Some Like It Hot", *Mother Jones*, http://motherjones.com/environment/2005/05/some-it-hot, accessed 2 July 2009.

MORTON, ADAM (2009) "The Sceptic's Shadow of Doubt", *The Age*, 2 May, p. 4.

MUGGLI, MONIQUE E., HURT, RICHARD D. and BECKER, LEE B. (2004) "Turning Free Speech into Corporate Speech: Philip Morris' efforts to influence U.S. and European journalists regarding the U.S. EPA report on secondhand smoke", *Preventive Medicine* 39, pp. 568–80.

NAHAN, MIKE (2003) "The Demise of Science", *The IPA Review* 55(4), p. 2.

NELKIN, DOROTHY (1987) *Selling Science: how the press covers science and technology*, New York: W. H. Freeman.

NELSON, ROBERT H. (1996) "How Much Is God Worth? The problems—economic and theological—of existence value", http://cei.org/studies-issue-analysis/how-much-god-worth, accessed 23 November 2010.

ORESKES, NAOMI and CONWAY, ERIC M. (2010) *Merchants of Doubt: how a handful of scientists obscured the truth on issues from tobacco smoke to global warming*, New York: Bloomsbury Press.

PAINTER, JAMES (2011) *Poles Apart: the international reporting of climate scepticism*, Oxford: Reuters Institute for the Study of Journalism, University of Oxford.

PEARSE, GUY (2007) *High and Dry: John Howard, climate change and the selling of Australia's future*, Camberwell: Viking/Penguin Books.

PEARSE, GUY (2009) "Quarry Vision: coal, climate change and the end of the resources boom", *Australian Quarterly Essay*, No. 33, Melbourne: Black Inc.

PEARSON, CHRISTOPHER (2009) "Sceptic Spells Doom for Alarmist Religion", *The Weekend Australian*, 18 April, p. 26.

PLIMER, IAN (2009a) *Heaven and Earth: global warming—the missing science*, Ballan: Connor Court Publishing.

PLIMER, IAN (2009b) "Hot-air Doomsayers", *The Australian*, *5* May, p. 10.

RAMPTON, SHELDON and STAUBER, JOHN (2001) *Trust Us, We're Experts! How industry manipulates science and gambles with your future*, New York: Penguin Putnam.

ROSKAM JOHN (2010) "An Open Letter to All Parliamentarians", http://www.ipa.org.au/sectors/climate-change/publication/1798/an-open-letter-to-all-parliamentarians/pg/3, accessed 18 December 2011.

SANDIFORD, MIKE (2009) "Geology Points to Dangers Ahead", *The Australian*, 6 May, p. 16.

SCHLESINGER, PHILIP (1990) "Rethinking the Sociology of Journalism: source strategies and the limits of media-centrism", in: Marjorie Ferguson (Ed.), *Public Communication: the new imperatives: future directions for media research*, London: Sage.

SMITH, MARCUS and MARDEN, PETER (2008) "Conservative Think Tanks and Public Politics", *Australian Journal of Political Science* 43(2), pp. 699–717.

SOLEY, LAWRENCE C. (1993) *The News Shapers: the sources who explain the news*, New York: Greenwood Publishing.

STONE, DIANE (1996) *Capturing the Political Imagination: think tanks and the policy process*, London: Routledge.

TAYLOR, MARIA (2010) "Perspectives on Framing and Climate Change in Australia 1987–2001", in: *Proceedings of the Environmental Research Event 2010*, Rockhampton: CQUniversity.

THE AUSTRALIAN (2009a) "Global Warming Beat-ups Threaten Scientific Credibility", *The Australian*, 9 April, p. 13.

THE AUSTRALIAN (2009b) "More Heat Than Light", The Weekend Australian, 18 April, p 16.

THE CANBERRA TIMES (2005) "Policy Tampering: it's a scandal that has scientists fuming", 13 June, p. 4.

UNION OF CONCERNED SCIENTISTS (2007) "Smoke, Mirrors and Hot Air: how ExxonMobil uses big tobacco's tactics to manufacture uncertainty on climate science", www.ucsusa.org/assets/documents/global_warming/exxon_report.pdf, accessed 13 October 2011.

VERON, JOHN E. N. (2009) "Tricky Dance Trips Up on Climate Facts", *The Canberra Times*, 9 May.

VINCENT, PETER (2004) "False Profits", *The Sydney Morning Herald*, 4 March, p. 6.

WALKER, JAMIE (2009) "The Climate of Consensus Is Precipitating a Disaster for Science", *The Weekend Australian*, 18 April, p. 18.

WALTER, MALCOLM (2009) "Heaven and Earth—review by Malcolm Walter", *The Science Show*, Australian Broadcasting Corporation, Radio National, 6 June, http://www.abc.net.au/rn/scienceshow/stories/2009/2586947.htm, accessed 8 September 2009.

WILSON, KRIS (1995) "Mass Media as Sources of Global Warming Knowledge", *Mass Communication Review* 22(1/2), pp. 75–89.

WILSON, KRIS (2000) "Communicating Climate Change Through the Media", in: Stuart Allan, Barbara Adam and C.ynthia Carter (Eds), *Environmental Risks and the Media*, London: Routledge.

WOODROFFE, COLIN, ENGLAND, MATTHEW, PEARMAN, GRAEME, BINDOFF, NATHAN, BROOK, BARRY, MARCHANT, HARVEY and LOWE, IAN (2009) "Rapid Roundup: new book by Ian Plimer doubts human-induced climate change—experts respond", Australian Science Media Centre, http://www.aussmc.org/2009/04/rapid-roundup-new-book-by-ian-plimer-doubts-human-induced-climate-change-experts-respond/, accessed 8 September 2009.

Index

Page numbers in **bold** type refer to figures
Page numbers in *italic* type refer to tables
Page number followed by 'n' refer to notes

www.routledge.com/9780415622899

ROUTLEDGE

Related titles from Routledge

Foreign Correspondence

Edited by Maxwell John Hamilton and Regina G. Lawrence

Despite the importance of foreign news, its history, transformation and indeed its future have not been much studied. The need to redress this neglect and the desire to assess the impact of new media technologies on the future of journalism, including foreign correspondence, provide the motivation for this stimulating, exciting and thought-provoking book.

While the old economic models supporting news have crumbled in the wake of new media technologies, these changes have the potential to bring new and improved ways to inform people of foreign news. Journalism is being transformed by the effortlessly quick sharing of information across national boundaries. As such, we need to reconsider foreign correspondence and explore where such reporting is headed. This book discusses the current state and future prospects for foreign correspondence across the full range of media platforms, and assesses developments in the reporting of overseas news for audiences, governments and foreign policy in both contemporary and historical settings around the globe.

This book was originally published as a special issue of *Journalism Studies*.

May 2012: 246 x174: 160 pp
Hb: 978-0-415-62289-9
£85 / $135

For more information and to order a copy visit
www.routledge.com/9780415622899

Available from all good bookshops

www.routledge.com/9780415622905

ROUTLEDGE

Related titles from Routledge

How Journalism Uses History

Edited by Martin Conboy

How Journalism Uses History examines the various ways in which journalism uses history and historical sources in order to better understand the relationships between journalists, historians and journalism scholars. It highlights the ambiguous overlap between the role of the historian and that of the journalist, and underlines that there no longer seems to be reason to accept that one begins only where the other ends.

With Journalism Studies as a developing subject area throughout the world, journalism history is becoming a particularly vivacious field. As such, this book argues that, if historical study of this kind is to achieve its full potential, there needs to be a fuller and more consistent engagement with other academics studying the past.

The book's contributors discuss the core themes which inform history's relationship with journalism from a wide range of geographical and methodological perspectives. They aim to create more ambitious conversations about using journalism both as a source for understanding the past, and for clarifying ideas about its role as constituent of the public sphere in using discourse and tradition to connect contemporary audiences with history.

This book was originally published as a special issue of *Journalism Practice*.

May 2012: 246 x 174: 136pp
Hb: 978-0-415-622905
£85 / $135

For more information and to order a copy visit
www.routledge.com/9780415622905

Available from all good bookshops

www.routledge.com/9780415622851

ROUTLEDGE

Related titles from Routledge

Explorations in Global Media Ethics

Edited by Muhammad Ayish and Shakuntala Rao

Studies of global media and journalism have repeatedly returned to discussions of ethics. This book highlights the difficulty that journalists encounter when establishing appropriate ethical practices and marks the pressing importance of global media ethics as a subject of current debate. A wide range of contributors – both scholars and practitioners of journalism – identify how changes in journalism practice, developments in new media technologies, legal regulations, and shifting patterns of ownership all play a role in creating ethical tensions for journalists, with some chapters in the book suggesting practical solutions to this pertinent issue. The growing need to faithfully represent other diverse cultural groups is also considered.

Explorations in Global Media Ethics recognises that, with the escalation of globalisation and a public striving for honest quality media, journalists around the world face an increasing pressure to comply with and simultaneously satisfy diverse ethical practices at both a local and a global level. The book sympathises with the position of the journalist and calls for greater consideration of his ambiguous role.

This book was originally published as a special issue of *Journalism Studies*.

May 2012: 246 x 174: 124 pp
Hb: 978-0-415-62285-1
£85 / $135

For more information and to order a copy visit
www.routledge.com/9780415622851

Available from all good bookshops

JOURNALISM PRACTICE

Editor: **Bob Franklin,** *Cardiff University, UK*

Journalism Practice provides opportunities for reflective, critical and research-based studies focused on the professional practice of journalism. *Journalism Practice*'s primary concern is to analyse and explore issues of practice and professional relevance. The journal aims to complement current trends to expansion in the teaching and analysis of journalism practice within the academy, reflection on the emergence of a reflective curriculum and thereby help to consolidate journalism as an intellectual discipline within the landscape of higher education.

Journalism Practice is devoted to: the study and analysis of significant issues arising from journalism as a field of professional practice; relevant developments in journalism training and education, as well as the construction of a reflective curriculum for journalism; analysis of journalism practice across the distinctive but converging media platforms of magazines, newspapers, online, radio and television; and the provision of a public space for practice-led, scholarly contributions from journalists as well as academics.

www.tandfonline.com/rjop

Routledge
Taylor & Francis Group

JOURNALISM STUDIES

LISTED IN THE THOMSON REUTERS SOCIAL SCIENCES INDEX®

Editor: **Bob Franklin,** *Cardiff University, UK*

Journalism Studies is an international peer-reviewed journal, published by Routledge, Taylor & Francis, which provides a forum for the critical discussion and study of journalism as both a subject of academic inquiry and an arena of professional practice. The journal's editorial board and contributors reflect the intellectual interests of a global community of academics and practitioners concerned with addressing and analysing all aspects of journalism scholarship, journalism practice and journalism education.

Journalism Studies pursues an ambitious agenda which seeks to explore the widest possible range of media within which journalism is conducted (including multimedia), as well as analysing the full range of journalistic specialisms from sport and entertainment coverage to the central concerns of news, politics, current affairs, public relations and advertising.

www.tandfonline.com/rjos

Routledge
Taylor & Francis Group